BRUGSCH'S 1883 THESAURUS

Volume I

TRANSLATED FROM GERMAN INTO ENGLISH
WITH ADDITIONAL NOTES

BY

JOHN M. BUNKER
KAREN L. PRESSLER

Bunker Pressler Books
2022

BRUGSCH'S 1883 THESAURUS

English Translation by John M. Bunker and Karen L. Pressler

Originally published in german as
THESAURUS INSCRIPTIONUM AEGYPTIACARUM
by Heinrich Brugsch

ISBN: 978-1-7325792-7-9

Bunker Pressler Books

"Two and a Guide"

Questions regarding this book can be addressed to:

Bunker Pressler Books
8829 Heffelfinger Rd.
Churubusco, IN 46723

Email:
Bunker.Pressler@gmail.com

Website:
https://sites.google.com/site/edgarcayceandthehallofrecords/

THESAURUS

INSCRIPTIONUM AEGYPTIACARUM

ANCIENT EGYPTIAN INSCRIPTIONS

COLLECTED, COMPARED, TRANSLATED, EXPLAINED AND AUTHORIZED

BY

HEINRICH BRUGSCH

PART ONE

LEIPZIG 1883.

J. C. HINRICHS'SCHE BOOKSHOP

Translator's Introduction

This is a translation of the first part of Brugsch's *Thesaurus Inscriptionum Aegyptiacarum*, a two-part work originally published in one volume. The translation is made in an effort to make one of the early important works in Egyptology available to the English-speaking world.

The translation presented many difficulties, one being the fact that the original document was handwritten. Brugsch also opted to use abbreviations freely, many of which are uncommon and challenging to interpret. However, the translators are quite familiar with the subject matter, which was a helpful advantage in creating this English edition.

Many of the charts and tables from the original edition have been copied directly, with German text translated to English; the ancient Egyptian and Greek are left as Brugsch has written them.

Brugsch used only a few footnotes, which are included in the translation. Additional notes were added by the translator when considered helpful to the reader, including the full titles for the abbreviated references given. These are distinguished in the footnotes with the designation "Translator note:" Otherwise, any footnotes are from the original Brugsch edition.

ASTRONOMICAL AND ASTROLOGICAL

INSCRIPTIONS

OLD EGYPTIAN MONUMENTS

COLLECTED, TRANSLATED AND AUTHORIZED

BY

HEINRICH BRUGSCH

LEIPZIG 1883.

INTRODUCTION.

The copies of the astronomical inscriptions, which have survived to the present day on the Egyptian monuments of older and more recent times, as far as they are accessible to the eye, form the first volume of the *Thesaurus Inscriptionnum Aegyptiacarum.* They have been copied by the editor on the spot during his long stay in Egypt, have been compared with texts that have already been published, and may deserve the claim to a complete compilation as well as to a correct reproduction of the existing astronomical material of the monuments. The connoisseurs of this part of Egyptian antiquity will not miss the new, partly unknown, partly little known and on the other hand unfortunately incorrectly published, which the editor gives in this work of science, with the good intention to fill a large and sensitive gap in Egyptian antiquity. The transcriptions, translations and explanations attached to the inscriptions will also give those who are more distant from the ancient Egyptian decipherments the opportunity to familiarize themselves with the rich content of the astronomical doctrines and views of the ancient Egyptians, and give the ancient Egyptian traditions their proper scientific value with the help of their own knowledge in the field of modern astronomy. Following the example of the monuments, the editor has given preference to the tabular arrangement of the traditions, in descending chronological order, over the merely descriptive representation, and has kept the more deeply founded differences of the astronomical doctrines of the older epoch until the end of the Pharaonic empire and the younger of the Greco-Roman time as strictly as possible from each other. Only in this way it seemed to him to be possible to facilitate the overview of the collected astronomical inscriptions for the connoisseur and non-connoisseur of the Egyptian scripts and render the most essential services to the understanding of them.

Quite apart from the extraordinary interest that is historically linked to the astronomical knowledge of the oldest cultured people on earth, the earliest evidence of which, almost a thousand years before the times of the beginnings of Greek astronomy, goes back to the epoch of the brilliant nineteenth (Theban) dynasty of the pharaonic kingdom, - the understanding of those ancient Egyptian traditions, in their progressive development from the astronomical point of view provides the only means of determining several immovably fixed points of ancient Egyptian history on the basis of astronomical calculations, thereby gaining the basis for not just manethonic but an exact chronology. In his introduction to *Chronologie der Aegypter*

(1849), Lepsius first demonstrated the astronomical basis of the same in a learned and astute manner with the aid of the monumental traditions. The critical results of his studies form the firm point of departure for all present and future investigations in the specified field and have paved the way for the careful researcher to proceed further. The increasing knowledge of the ancient Egyptian language and script decipherment, the discovery of important unknown monuments and occasional corrections and additions through the publication of already known astronomical texts have in detail corrected and completed the results of research gained by the old master of modern Egyptology, but have not been able to finish the outlines of the overall picture, the lines of which have been drawn by him once and for all with a firm and sure hand. The present "astronomical inscriptions" are also to be regarded as parts of that overall picture in their cohesive arrangement. They fill in open gaps and add details that had escaped the sharp eye of the founder of ancient Egyptian chronology or remained unknown to him.

As the editor had set himself the task to arrange all of the astronomical material of the monuments in a clear way, to explain it from the philological point of view and to arrange it in groups. He has abstained from any judgment in the astronomical sense due to the lack of astronomical knowledge as is available to the astronomer, and has left it to the experts to use what is offered for the benefit of the historical research. Just as misunderstood inscriptions and representations, especially in earlier times, have been shown to have led the most learned and astute astronomers to the most unsuccessful, if laborious, calculations, for want of an accurate philological interpretation of the texts, so, on the other hand, there is no lack of examples in which astronomically untrained Egyptologists have wrested the most deceptive data from the astronomical traditions of the monuments.

The numerous inscriptions and illustrations, which fill the following pages, are personally written down and drawn by the author. Little practiced in ancient Egyptian painting, he preferred to use published models for several illustrations, of the kind that science owes to the master hand of the painter E. Weidenbach (in the monuments of the Prussian expedition and in the murals of the Berlin Museum). Of course, the drawings have been compared with his own copies beforehand and, where necessary, improved in the relevant places. Also the publications of the astronomical pictures on the ceilings of the temples of Dendera, Esne and Edfu, which are in the *Description de l'Egypte*, have also given the editor the opportunity to use the ceiling images of the mentioned sites, which are currently barely accessible to the eye, as a control. The astronomical calculations and determinations, which the editor believes are based on an erroneous interpretation of astronomical inscriptions, have been passed over by him with silence, since the confirmation seemed to him more useful and more pleasant than the refutation of unfortunate or unscientific theories.

Charlottenburg, November 26, 1882.

H. B.

Table of Contents

Mr. Hermann Rost

Knight of the K. Prussia Order of the Crown

Officer of the K. Ottoman Medjidic Order

owner of the K.K. Austria large gold medal "literis et artibus"

owner of the J.C. Hinrichs' bookshop

in Leipzig

consecrates this work

Thesaurus Inscriptionum

Aegyptiacarum

as an expression of his joy for

more than twenty five years of faithful connection

between publisher and author

in

the most grateful spirit

Heinrich Brugsch

The Astronomical Inscriptions on the ceiling in the pronaos of the temple of Dendera from the time of the Emperor Tiberius

Although the figurative representations in brightly painted skin reliefs, which are found on the ceiling of the pronaos mentioned above, have become more generally known through the publication in Volume IV of the *Descripton de l'Egypte* (Antiquites) - partly with errors in the sequence of the figures and with mistakes in the details - have formed the subject of numerous scholarly investigations in the field of ancient Egyptian astronomy and calculating chronology: nevertheless, the inscriptions attached to the individual figures, which are so necessary for the understanding of the representations, have not been made accessible to science in their entire context from any side until now. Only particulars, such as the names of the hours and the decans, have been published in scattered publications. The extremely laborious transcription of the following texts could only be achieved with the help of sharp glasses and with the best possible illuminating sunshine. A repeated search and comparison of the copies taken should provide sufficient security for their reliability and accuracy. To make it easier to understand the arrangement of the whole and the distribution of the individual, I am following the outline of the "*Description de l'Egypte*" on a reduced scale, with the plan of the ceiling in the drawing below (pages 2 and 3). In general, I note in advance that the three columns A, B, C with their figures and texts belonged to the course of the sun and the constellations of the southern sky, while the three columns A', B', C' refer to the lunar sphere and the constellations of the northern sky. This preliminary reference will be quite sufficient to facilitate the understanding of the overall conception. In order to save space, the figurative representations are in most cases left aside in the following, if the descriptions of them correspond faithfully to the originals and allow them to be clearly identified without misunderstanding. Wherever it was a matter of an essential confirmation or the correction of real errors, the graphic representation on my part was added to the inscription.

The unknown author of that important ceiling ornament was able to present in pictures and words the astronomical-calendrical ideas and the core knowledge of his time and his country for the Egyptian eye vividly and clearly (not without the influence of Greek-Roman views, especially demonstrable in the introduction of two signs of the animal circle into the Egyptian sphere), so also for modern science that monument of Egyptian wisdom in the field of astronomy, together with the yearly form of the Egyptian calendar, has the high importance of a fixed starting point. And also in individual traditions of an astronomical nature in the temples and kings tombs at Thebes from the times of the XIX and XX dynasties, which occasionally become apparent through a more precise implementation of the treated material in detail, they still lack the transparent connection and the clearer understanding of the astronomical ceiling paintings of the Pronaos of Dendera. After this preliminary remark, I will move on from strip to strip to the inscriptions themselves.

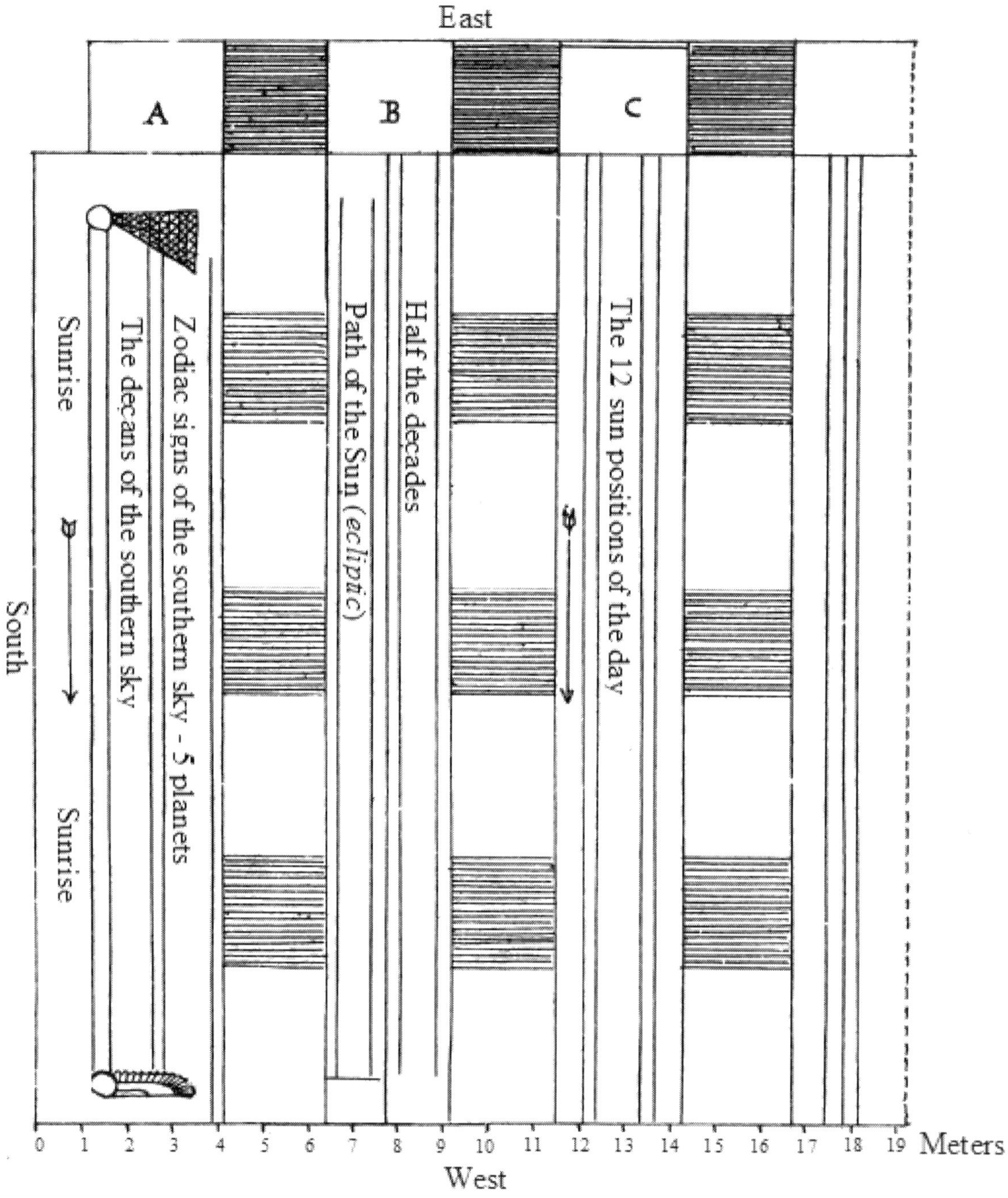
East
A
B
C
Zodiac signs of the southern sky - 5 planets
The decans of the southern sky
Sunrise
Sunrise
South
Path of the Sun (ecliptic)
Half the decades
The 12 sun positions of the day
0 1 2 3 4 5 6 7 8 9 10 11 12 13 14 15 16 17 18 19 Meters
West

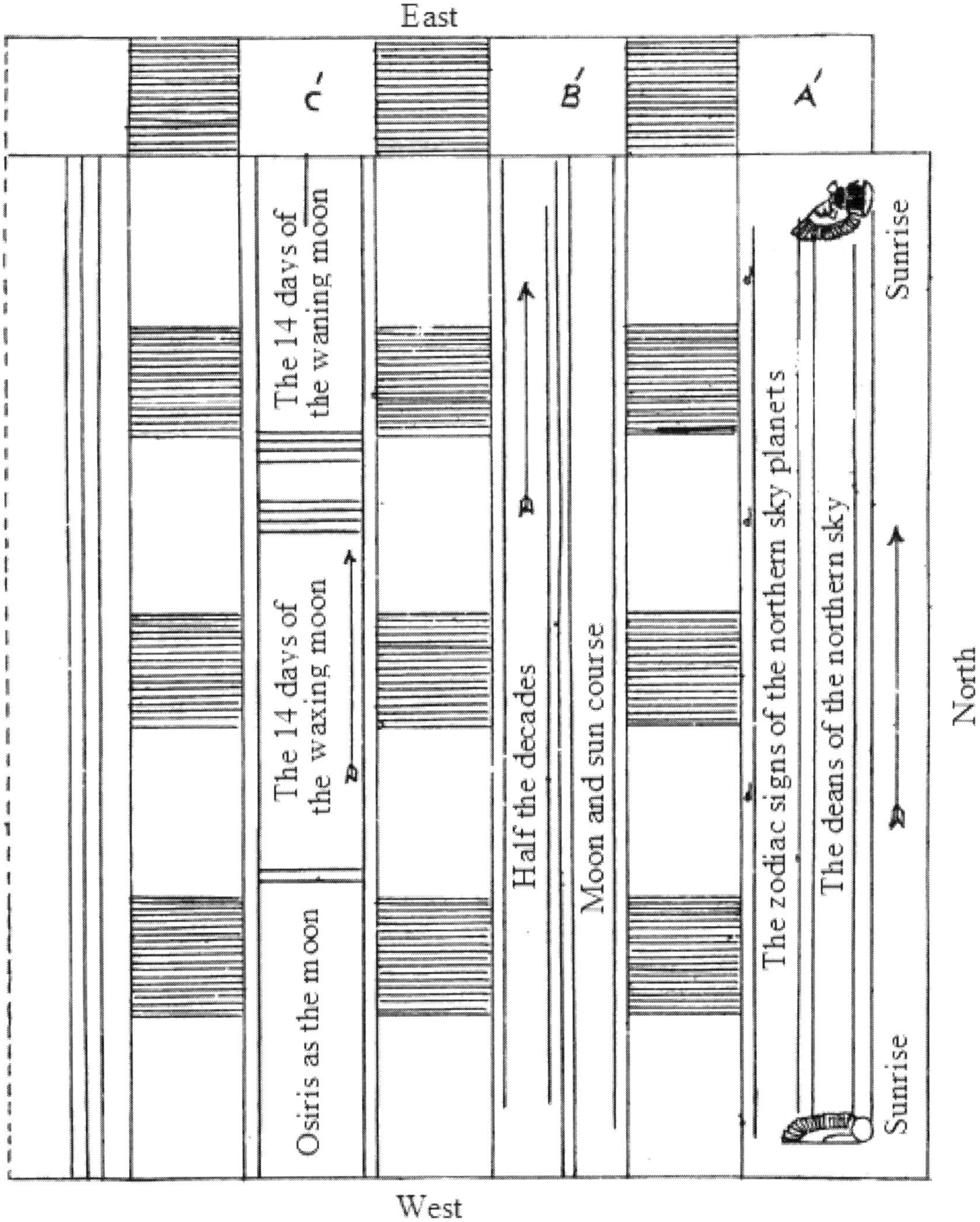
East
C′
B′
A′
The 14 days of the waning moon
The 14 days of the waxing moon
Osiris as the moon
Half the decades
Moon and sun course
The zodiac signs of the northern sky planets
The deans of the northern sky
Sunrise
Sunrise
North
West

North Side, Strip A', Upper Representation, Starting from the East

1. the goddess 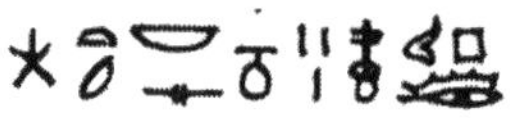 pir-nofru-rn-neb-set ("who shrieks the glories of her lord," i.e. the sun god) of the twelfth hour of the night.

2. the zodiacal sign of the lion.

3. destroyed image.

4.  astronomical sign of unknown meaning.

5. the goddess 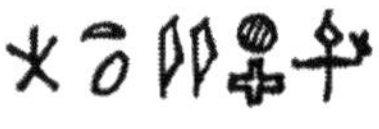 xesef-xemit ("Defender of the Stranger") of the eleventh hour of the night.

6. the goddess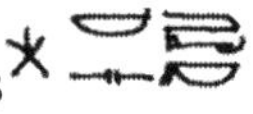 mak-neb-es ("her lord's guardian") at the tenth hour of the night.

7. the zodiacal sign of the young woman

8. the planet 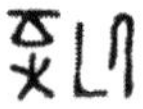 sebek (Mercury) represented as .

9. bull-headed deity carrying the instrument  in his hand.

10. the goddess 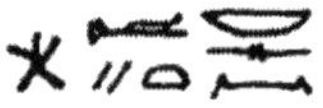 neb-sendti ("Lady of Terror") of the ninth hour of the night.

11. the goddess  semer-nesr-et the eighth hour of the night.

12. the zodiacal sign of the scales.

13. figure enclosed in a disk: , to the left of it:

14. the planet 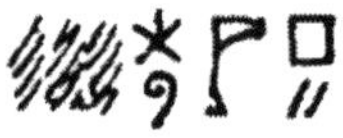 pi-nuter-du(a) "the god of the morning" i.e. Venus, represented as . Subsequent to this:

15. and 16. the goddesses of the seventh and sixth hour of the night. Both names are destroyed and illegible on the strip.

17. picture of a standing hippopotamus, the crown  on the human head, carrying two 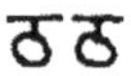 in the hands.

18. the zodiacal sign of the Scorpion. To the left above it:

19. a nameless planet 

20. the constellation , jackal on the instrument 

21. the goddess  *neb-anx* ("mistress of life") of the *fifth* hour of the night.

22. the goddess 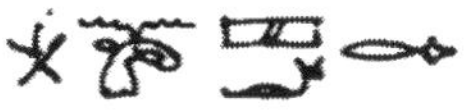 *a sef* ("very handsome") the *fourth* hour of the night.

23. the Zodiac sign of *Sagittarius*, above it to the left:

24. a planet in the form: 

25. the goddess 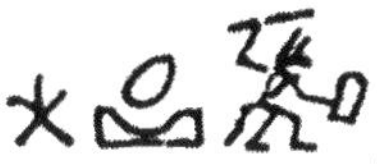 *seheru du* ("scarer of evil") the *third* hour of the night.

26. Image of the god Horus, referred to as  *Hru Xer setebu* "Horus fighter of the enemy", piercing with a spear

27. the bull's thigh in the northern sky or the constellation of the *Great Bear* tied with a chain,

28. the hippopotamus goddess ( *reret*) holding firmly in the right hand

The entire illustration, referring to the *turn of winter*, is shown in the following illustration:

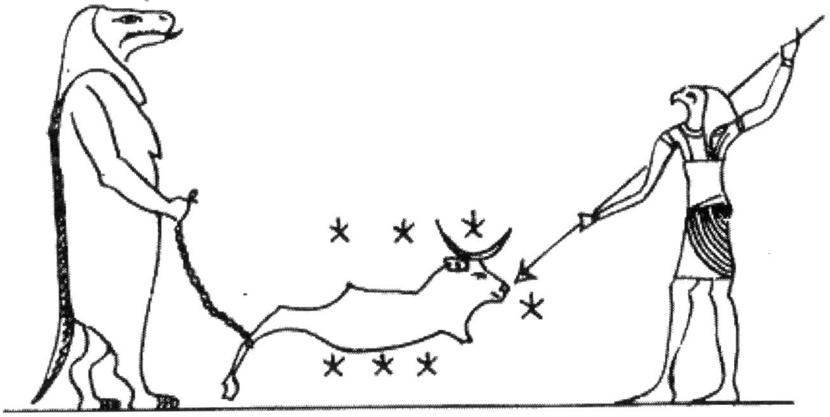

The seven stars of the *Great Bear*[1]

29. the goddess  *sar-neb-es* ("her Lord's ruler") on the *second* hour of the night, behind which appears:

30. the planet figure  (sparrowhawk with bull's head) of *Saturn*.[2]

31. the Zodical sign of Capricorn. At the end:

32. the goddess of the *first* hour of the night 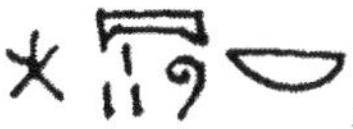 *neb-hru* (sic).

---

[1] Translator note: Brusch's understanding of this image is mistaken. The constellation represented is Taurus, not the Great Bear.

[2] Translator note: The planet figure is Jupiter, not Saturn.

The continuation of the above images and texts provides the

South Side, Strip A, Upper Representation, Starting from the West

1. the goddess of the *first* hour of the night  *neb-hru-t.*

2. the bull-headed shape of the planet *Saturn,*[3] in the inscription called  *Hru-ka* "Horus-bull."

3. god with vulture's head standing on a goose .

4. god, an *ankh* [4] in his right hand, holding an antelope (Typhonian animal) as a sacrifice with his left.[5]

5. headless man in the position:  , without inscription

6. the goddess of the *second* hour of the night 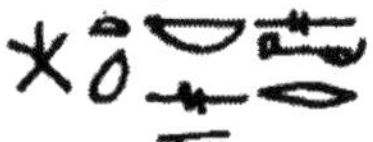 *sar-neb-set..*

7. the goddess of the *third* hour of the night 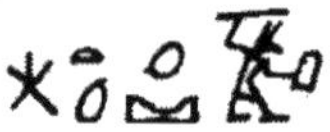 *seheru-du-t.*

8. the zodiacal sign of Aquarius.

9. the planet Mars depicted as a sparrow-headed god () with the inscription 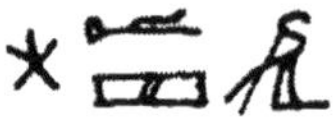 *Hru-dos* "Horus the Red".

10. the goddess of the *fourth* hour of the night  *a-sef-t.*

11. the zodiacal sign of *Pisces.*

12. the goddess of the *fifth* hour of the night  *neb-anx.*

13. in a disk, a man holding a pig by the tail (*xes-deb*).

14. the planet *Jupiter* in the form of Horus  with the inscription  *Hru-up-sefa*, containing its Egyptian name.

15. the goddess of the *sixth* hour of the night 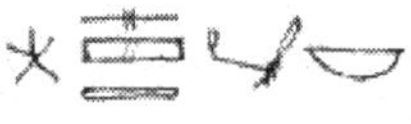 *neb-sor-sesta.*

16. that of the *seventh* hour of the night 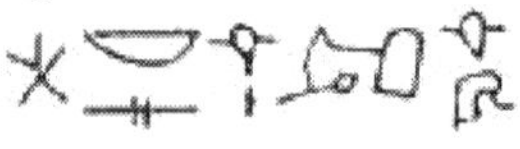 *hr-tp-xer-hr-neb-s.*

17. the zodiac sign of *Aries.*

18. lion headed god with scepter  in hand.

19. god with a star  on his head and  in his hand.

20. dog-headed monkey and antelope with their backs turned toward each other. On the head of the former animal .[6]

---

[3] Translator note: Should be planet Jupiter

[4] Translator note: Brugsch identified the *ankh* as a *messr*-knife.

[5] Translator note: This is the constellation Andromeda. (*see*: Bunker-Pressler. *The New Order of Things*, chapter 5, Dendera, to be published in 202.)

21. the planet *Venus*, shown with a double head:  with the addition: 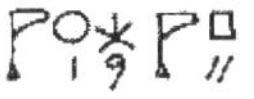 *pi-nuter-du* “the god of the morning”.

22. the goddess of the *eighth* hour of the night  ..

23. the goddess 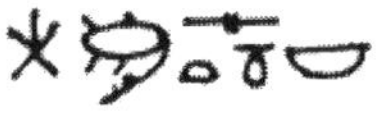 *neb sent* of the *ninth* hour of the night.
24. the zodiacal sign of the bull, on its back the disc of the new moon.

25. image of a deity with the plumage of the god *Anhur*:  (or ) on her head.

26. the planet *Mercury* represented as  with the inscription  *sebek* (see the planet table below).

27. the goddess 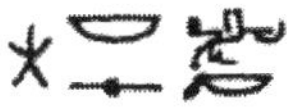 *mak-neb-s* at the *tenth* hour of the night.

28. the goddess 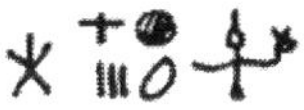 *xesef xamu* at the *eleventh* hour of the night.
29. the zodiacal sign of Gemini, represented by the dual deities *Su* and *Tafnut*.

30. the goddess  *par-neferu-en-neb-set*[7] of the *twelfth* hour of the night.

31. the constellation of *Sahu* or Orion: the inscription above Orion reads

 *sahu bi nuter seps en usiri*, i.e., Orion, the splendid godly soul of *Osiris*.

32. next picture:
(Horus on the papyrus column)
Summer Solstice (?)

33. *Sirius* is represented as a star image  by a cow lying in a ship, above it the text:

(1) “The divine *Sopedet* (Sothis), the great, the mistress of the year of repentance, the daughter of *Ra*, Isis, the mistress of heaven,
(2) currently rising to open a happy year, she passes peacefully behind her brother
(3) to the god as *Sahu* star (Orion) for son Horus (appears) as the sun, forever.”

34. the same star under the image of an upright goddess with the inscription 

---

[6] Translator note: Constellation Auriga and Shepherd’s Star Capella. (*see*: Bunker-Pressler. *The New Order of Things*, chapter 5, to be published in 2023.)

[7] Budge, *Gods of the Egyptians,* Volume 1, page 301, hour XII:  *par-neferu-en-neb-set.*

Brusch wrote: the goddess  *pir-[nofru-nu-nrb-]s* of the wolves.

"the divine <u>*Sothis*</u>. She is (like <u>*Satit*</u>, the goddess of the Nile flood)  in a ship. Behind her as a companion:

35. the goddess  <u>*anqet*</u>  pouring water out of cans to indicate the incoming flood.
36. the zodiacal sign of <u>*Cancer*</u> (Egyptian the  beetle) behind the morning sun rising over the temple.

<u>North Side, Strip A', Lower Representation, Starting from the East</u>

<u>The Decan Stars</u>

Behind the rising sun (in the shape of ) the following row of decan stars moving in barges:

1 .  <u>*phui-hru*</u> (in the case of Salmasius[8] φονόρ) Then God in the form of a boy before whom a serpent rises up , personification of the New Year's Day sun. Accordingly, he carries the designation 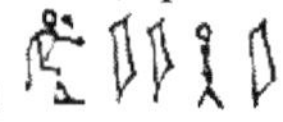 <u>*ahi*</u> "boy," by which the sun of the New Year was understood in the calendar texts of Dendera and Edfu. The Dendera calendar lists:

Month <u>*Thoth*</u>, day 2, at the fourth hour of the day (called <u>*Sestait*</u>) the emergence of the large lotus flowers in their symbolic perception as the greater god <u>*Ahi*</u> (sun of New Year's Day), son of the goddess Hathor. In Edfu (calendar I, column 1) the position referring to the first day of the Egyptian calendar year after a row of characters that have unfortunately been destroyed reads:

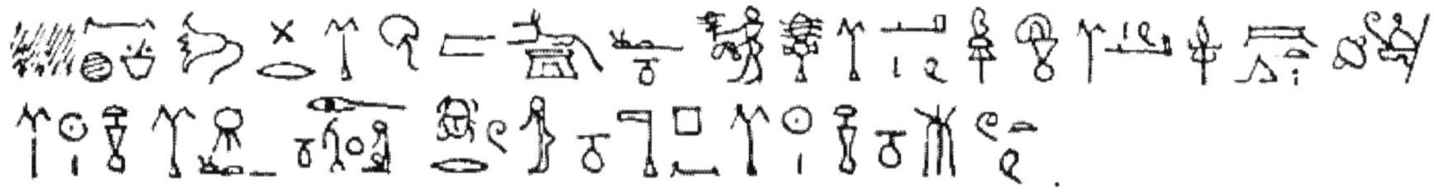

[New Year's Day: the mergence of the large lotus blossom in the form of a bud in its symbolic conception as god <u>*Ahi*</u>, who has a sistrum in his right hand and the symbol  (type of rosary) is lying in his left hand and his legs are spread. The calculation of (his) rule begins from the first day and from his rise. Osiris assumed the form of this god on the first day of his birth. In another text published in BHJ[9], LVIII, col.3 it is stated:

---

[8] <u>Translator note</u>: Claudius Salmasius 1588-1653, knew Hebrew, Arabic Syriac, Persian and Coptic, as well as Greek & Latin. Whenever Brugsch uses "Sal." He is probably referring to Claudius Salmasius. <u>Britannica.com/biography/Claudius-Salmasius.</u>

[9] <u>Translator note</u>: BHJ might refer to: Bergmann, Ernst von. *Hieroglyphische Inschriften*. (Brugsch used this abbreviation as such in his *Hieroglyphisch-demotisches Wörterbuch*, vol. 7.)

 "New Year's Day: the sun comes out of a lotus flower in the great sea". Frequent are the allusions to the lotus blossom in the great water, from which the sun child rises up to the sky in full light. After <u>*Ahi*</u> follows:

2. destroyed decan [to be completed by  <u>*knumm*</u>, Salmasius ΧΝΟΥΜΊϹ]

3. destroyed decan [read:  <u>*Xar-knum*</u>, Sal. χαρχνουμίϲ]

4.  <u>*ha-tat*</u>, Sal. ῾ΗΤΗΤ

5. 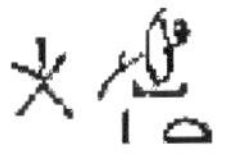 <u>*tat*</u>

6. 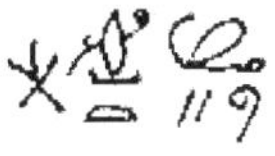 <u>*phui-tat*</u>, Sal. ΦΟΥΤΉΤ

7. 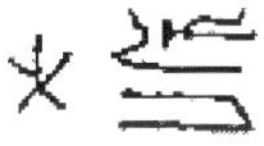 <u>*tomam*</u>, <u>*domam*</u>, <u>*dom*</u>, Sal. ΤΏΜ

8.  <u>*usta*</u>, Sal. ΟΥΕϹΤΕΒΙΚΏΤ

9. 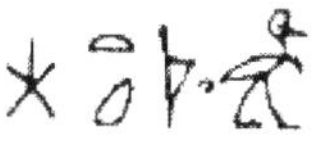 <u>*bikot*</u>

10. 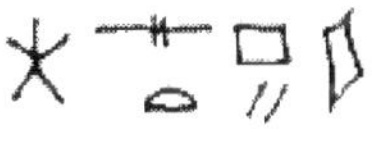 <u>*apiset*</u>, Sal. αφοϲό

11. 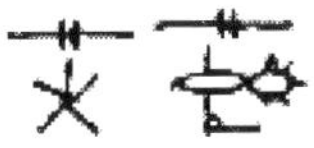 <u>*sebxas*</u>, Sal. ϹΟΥΧΏϹ

12.  <u>*apa-xont*</u>, Sal. ΤΠΗΧΌΝΤΙ

13.  <u>*hru ba ua*</u>, Sal. ῾ΡΗΟΥΏ

14.  <u>*sopt xon*</u>, Sal. ϹΠΤΧΝΈ

15. 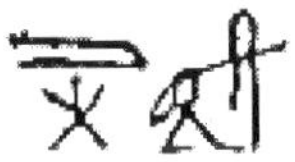 <u>*sesem*</u>, Sal. ϹΕϹΜΈ

16. 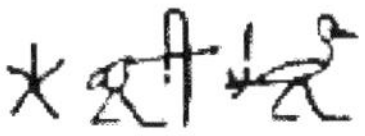 <u>*si sem*</u> Sal. ϹΙϹΕϹΜΈ

17. 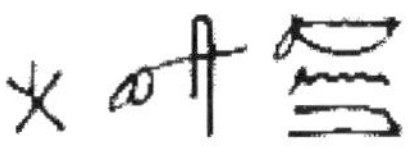 <u>*konem sesem*</u>, Sal. ΚΟΝΙΜΈ

18. 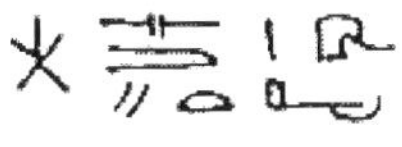 <u>*tpa-smati*</u>

This is followed by the picture of the setting sun  with the inscription 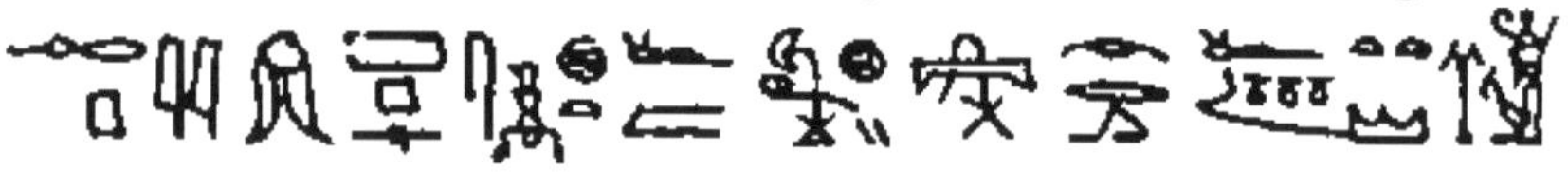 _api seps soxes-f em uxxi seper-f ma-nun em atum_, "The magnificent flyer descends in the evening. He comes to the west country (_Manun_) as god _Atum_."

[Continuation of the row of decans on the]

South Side, Strip A, Lower Representation, Starting from the West

Behind the setting sun, designated as: 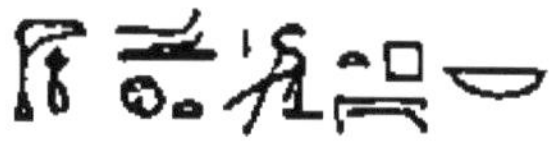 "Horus of Apollinopolis Magna, great god, Lord of Heaven, with the addition of 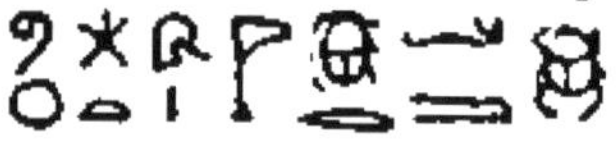 "He turns into a beetle every morning," follow the decans; which form the second half and the end of the previous kingdoms. At their head is a Horus god (): 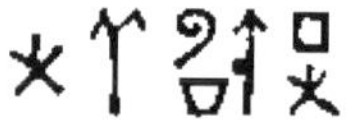 _pe-seb ua em seb_ "the evening star (_Hesperus_) as a star. (see Br. W.S.[10] p. 1358)

19. 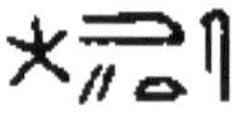 _smati_, Salmasius ϹΜΆΤ

20. 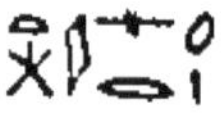 _si-srat_ Sal. ϹΙϹΡΏ

21.  _tpa-Xu_, Sal. ΤΠΗΧΎ

22.  _Xu_, Sal. ΧΎ

23. 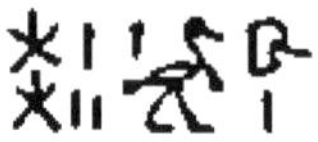 _tpa-bu_, Sal. ΤΠΙΒΙΟΎ

24. 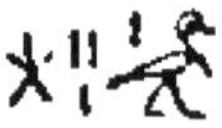 _biu_, Sal. ΒΊΟΥ

25.  _spa-biu_, Sal. ΤΠΗΒΙΟΎ

26.  _Xont-hru_, Sal. ΧΟΝΤΑΡΈ

27. 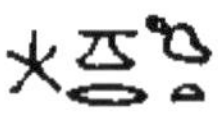 _xont-xer_, Sal. ΧΟΝΤΑΧΡΈ

28. 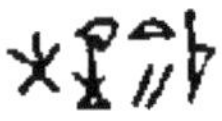 _Koti-Xo_, Goodwin ΚΑΤΚΟΥΆΤ

29. 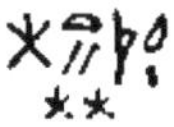 _Si-Koti_, Sal. ϹΙΚΈΤ

---

[10] Translator note: most likely Brugsch, _Worterbuch,_ supplement.

30. 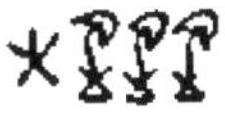 *Xou*, Sal. χώογ

31. 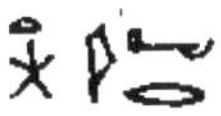 *arat*, Sal. ερω̄

32.  *remen-hru*, Sal. ρωμεναρέ

33.  *Θos-alk*, Sal. θοςόλκ

34.  *remen-xer*, Sal., Goodw. ρεμ(εν)χ(αρε)

35.  *uar*, Sal. ογάρς

This is followed by a ship with the deities Isis, Hathor and *Horsamta* (other designation of  see above) and in the very corner next to the eastern rising sun, which spreads its rays over the Hathor head (symbol of the temple of Tentyra), the boat:  with the heading: 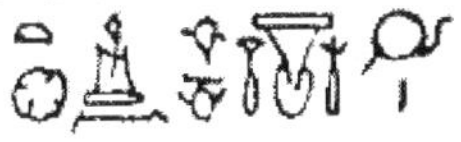 *ra-samta her-ab ant* "the sun, unifying the world, in Tentyra" as an expression of the new year.

In the margin a-a-a the following text. The translation of the inscription, which is addressed to the goddess Isis in her conception as Sothis star (Sirius), is as follows in the most literal translation:

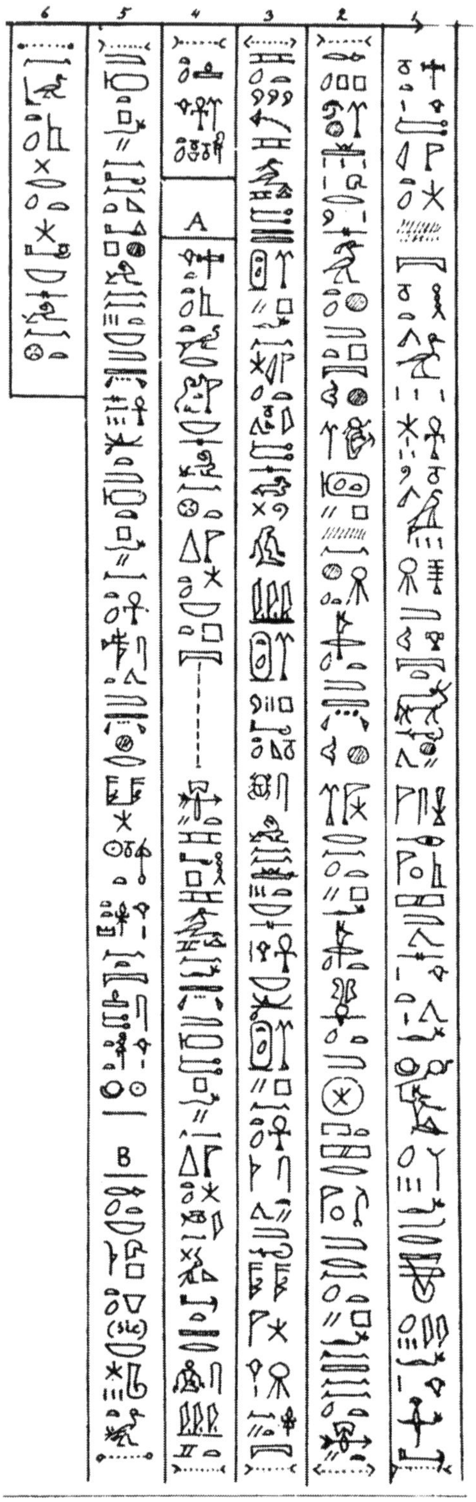

"Hail to you! Sothis star, Isis, the [mistress] of heaven, you queen of the rising souls (i.e. the decan constellations) of the gods. Radiant in the sky near her brother Osiris, she always walks in his footsteps, removing his enemy*) repelling the serpent Apophis (apop) with the glorious utterances of her mouth. You shine in the sky with the day god *Ra* in your name of *the shining one* (*xut*). You are mighty on earth with the earth god *Seb* (Kronos) in your name of *the mighty* (*usert*). You are great in depth in your name of *Tanent* (i.e. the great). You swell (*sati*) the Nile in your name of the divine *Sothis*. You embrace (*ankeu*) and make the field fertile in your name of the *Anket*. You create everything that is there, giving life (*anx*) to all people in your name of *Anxet* (life). Orbiting near the *Sahu* star (*Orion*) and rising in the east of the sky, you unite with life in the west of the sky. The inscription A, which is borrowed from the Sothis temple at Syene (ancient Egyptian *Suan*, today Aswan), repeats the content of the text from Dendera. The inscription B (see Col. 5-6), taken from the same Sothis sanctuary, describes the goddess as "the great one, the mistress of the beginning of the year, the queen and mistress of the decan stars, the daughter of the earth god *Seb* (Kronos), Isis the Great, the Mother of God, the Lady of Syene."

---

* Add: "together with his comrades"

North Side B' and South Side B, Strip B', Upper Representation, Starting from the East, Strip B from the West

Representation of the decades of the Egyptian_Year.

B' 1. Image of the wind god of the east (head destroyed).

2. Goddess with 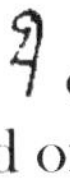 on her head. Inscription:  "... the great one, the daughter of the day god or the sun *Ra*."

3. Seated image of the god *Ra* with the inscription: 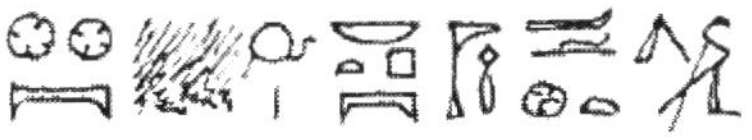 "the Apollinopolitan Horus, the great god, lord of heaven, the sun (of the rising in the?) sky."

4. Pregnant woman crouching on the birthing chair  with the caption  "the sky goddess (*nen*), giving birth to the sun."

5. Sphinx figure lying on a temple-like structure, referred to as "the divine figure of *Ra* above the box in the sphere of light (*Xut*)."

This is followed by the decades, personified as divine beings, according to their names and pictorial representations, comprising 24 subjects. Then at the end of them in the order from E-W as Figure:

29.  with the inscription: 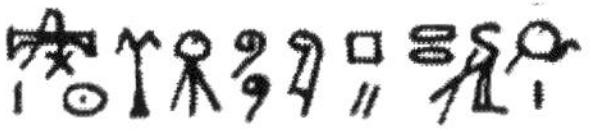 *Ra hru-xuti pi-suu em goh* "*Ra-Hor-Xuti* (the shining Horus), the ray of light in the night," the god conceived as the moon.

30.  called: 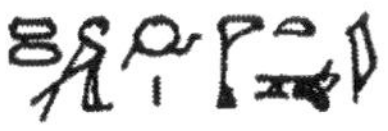 *Atum-Ra-Hor Xuti* "the evening, morning and noon sun."

31. the image of a reclining sphinx referred to as 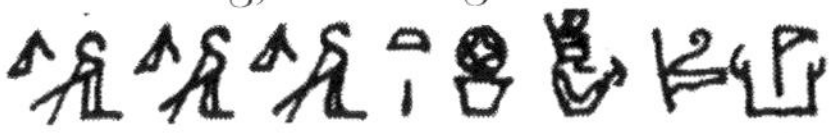 "the divine figure of *Atum* the father of gods."

32. the image of the pregnant woman on the birthing chair, called  "the Heavens as the bearer of *Atum*." Then,

33. the figure of the god 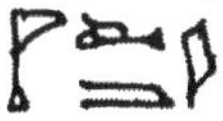 *Atum* enthroned as a king. At the end of the corridor the depiction of a wind man, who is referred to as the west wind by the inscription 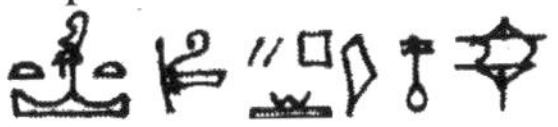 "the good wind from the direction of the west."

On the south side, strip B, there are also pictorial representations accompanying the decade names and figures. To the west appears the depiction of a wind god in flight, after the inscription 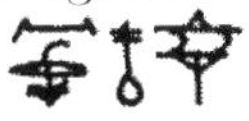 "*the good wind of the south*." To the east, immediately behind the last decade picture, two ships appear, each with a chapel inside. In the first throne *Horsamta* and behind him *Ra*, in the second *Isis-Hathor* and *Osiris*. A wind god concludes the whole performance. According to the general distribution, it corresponds to the north wind.

For the sake of completeness, the decade names and images of the pronaos with thoroughly related representations that deal with the same subject have been clearly put together, namely under B with the series already published by Mariette and Dumichen handed down in room xii of Dendera, under C with the list preserved in the Pronaos of Edfu, and finally under D with the decade figures present in the temple of Esne. The meaning of the metal and mineral names attached to the individual decades in Dendera (A', B') is difficult to see. The importance of the decade distribution for the overall presentation is indubitable and self-evident. The transcription and translation of the inscriptions added at the end of the following plates will allow the most accurate understanding of them.

A Dendera
B Dendera
C Edfu
A'
B'
A''
B''
C''
D'' Esna
sic
1 2 3 4 5 6 7 8 9 10 11

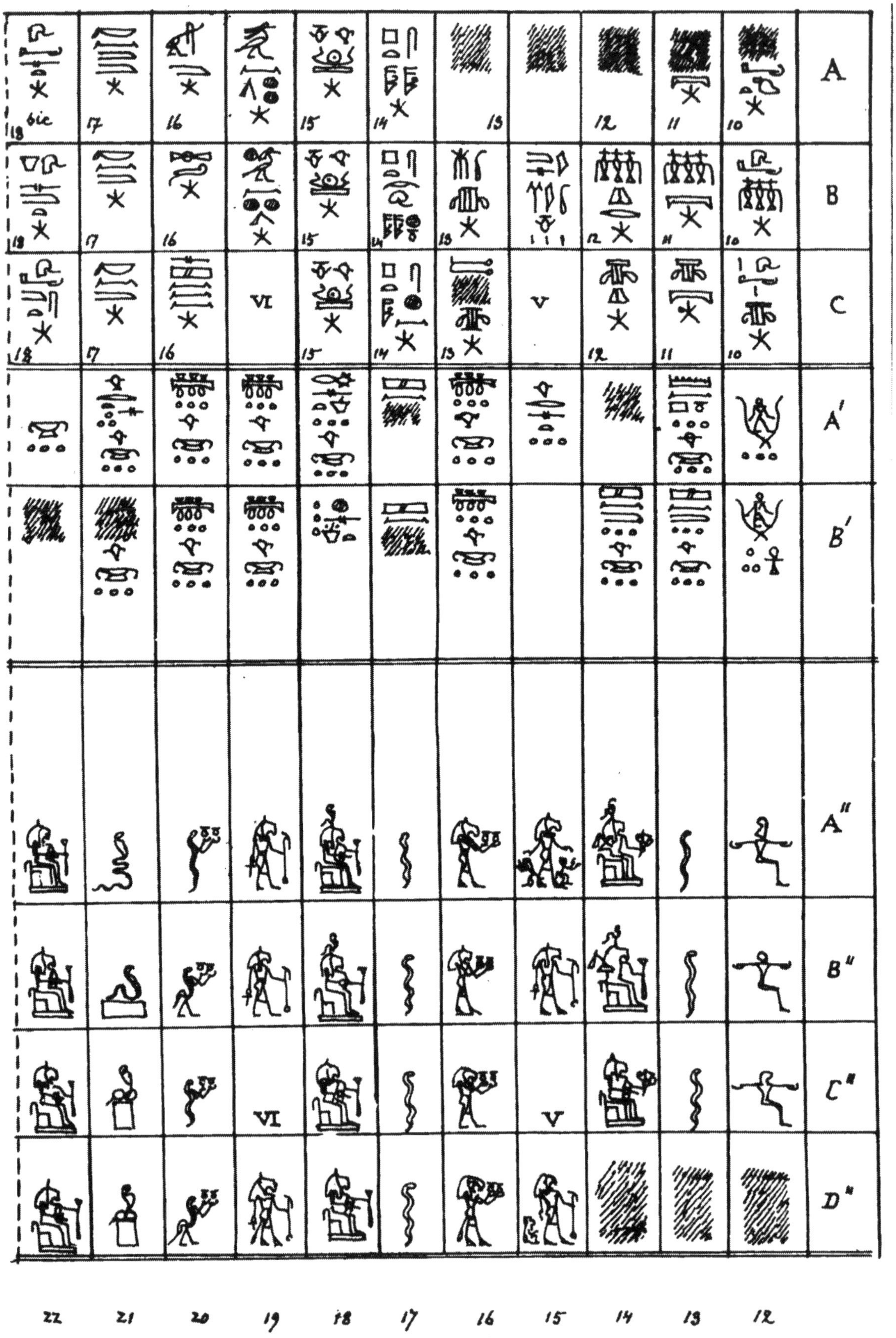
A
B
C
A'
B'
A''
B''
C''
D''
VI
V
22 21 20 19 18 17 16 15 14 13 12

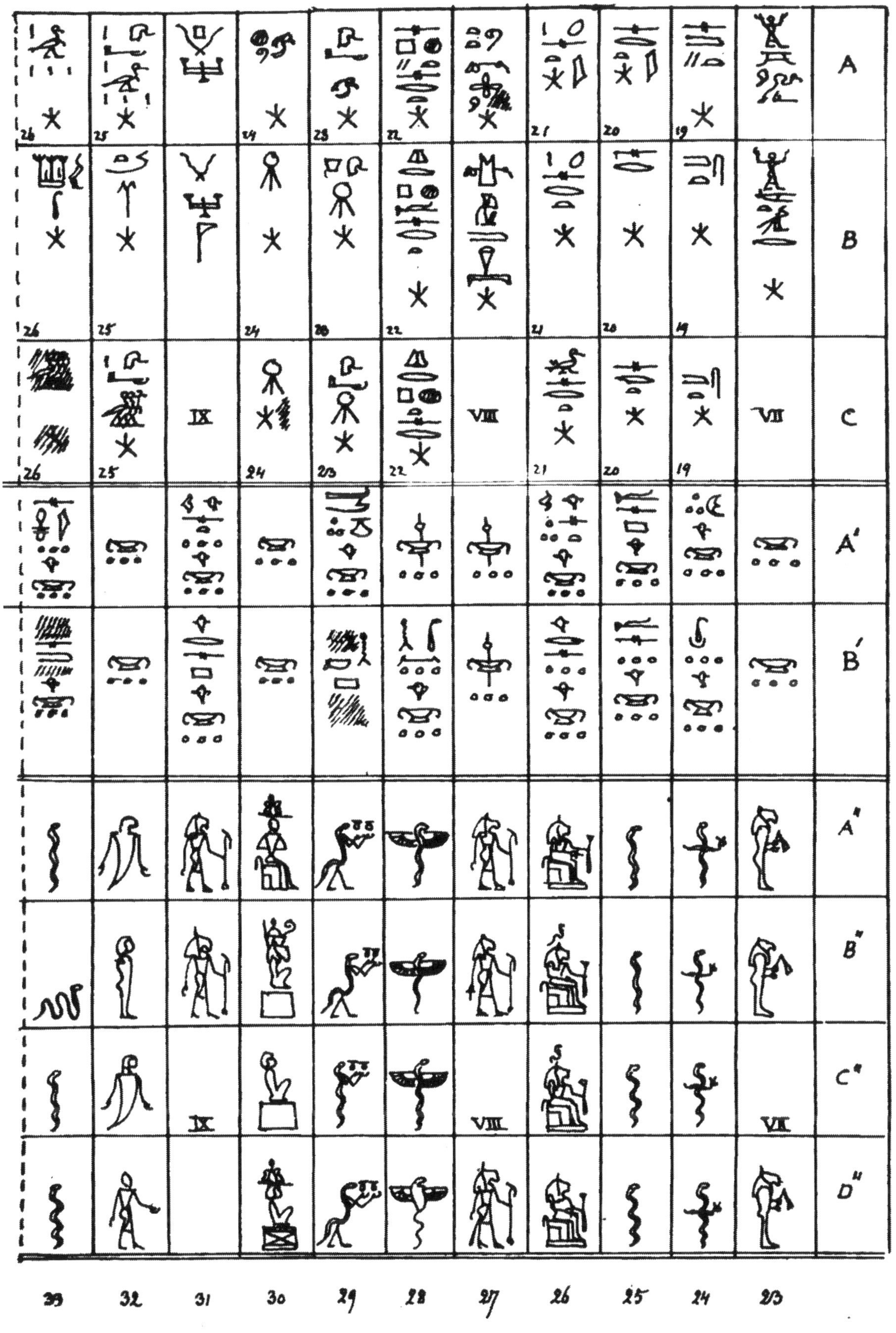
A
B
C
A'
B'
A''
B''
C''
D''
IX
VIII
VII
33 32 31 30 29 28 27 26 25 24 23

A
B
C
A'
B'
A"
B"
C"
D"
XII
XI
44 43 42 41 40 39 38 37 36 35 34

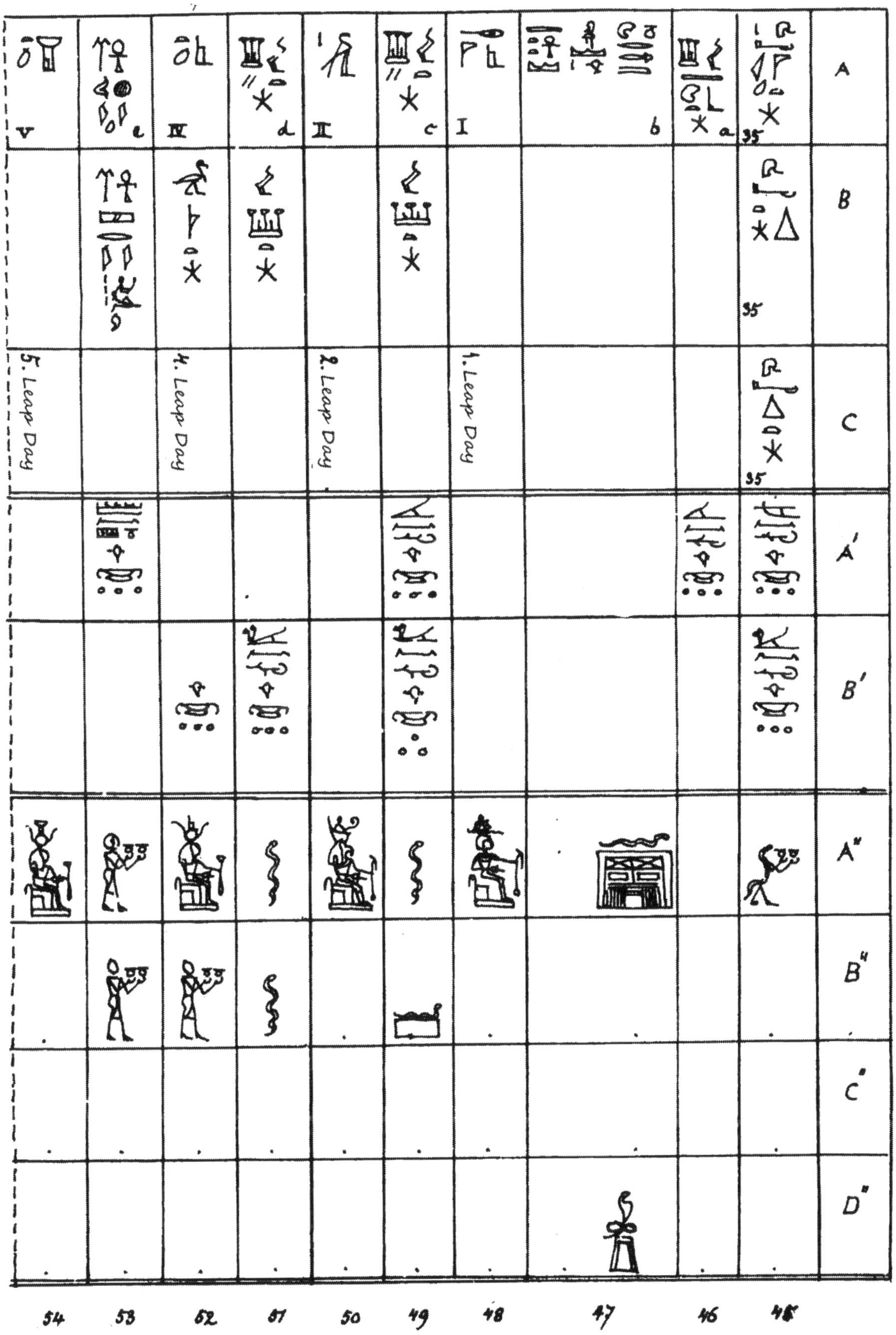
A
B
C
A'
B'
A''
B''
C''
D''
V
e
IV
d
II
c
I
b
a
35
5. Leap Day
4. Leap Day
2. Leap Day
1. Leap Day
54
53
52
51
50
49
48
47
46
45

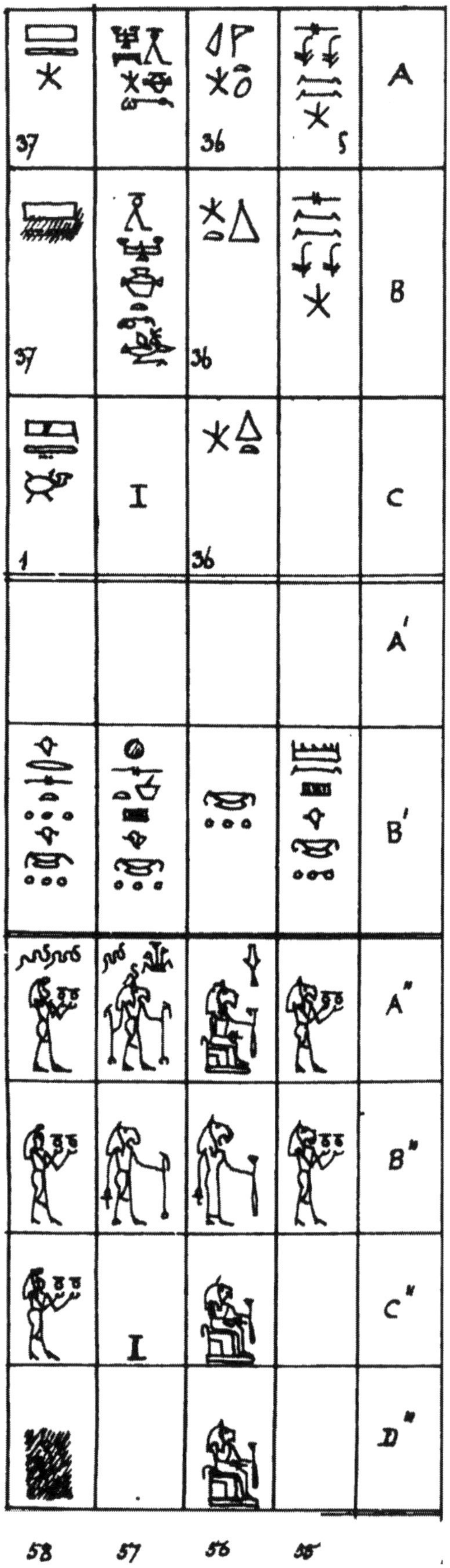

A comparative examination of these three lists, of which two (A, B), more precisely three (A", B", D"), belong to the Imperial period and one (C, C") to the Ptolemaic epoch, reveals important elements in the sequence of the decades, starting from their starting points, which are of great importance for the correct knowledge of the various forms of the year compared to the normal Sothis year.

Suffice it to say for the moment that, according to the Ptolemaic index from Edfu C, the decade named after the constellation of the Turtle (Greek. CIT) opens the year as a rule, while in Dendera it marks the end of the decades of the year. The pictures attached to the individual decades, despite some variations in detail, immediately reveal the intended systematic arrangement. Compare the following translation.

The Decades of the Ancient Egyptian Year according to the Indexes and Illustrations in Edfu and Dendera

№	Chief	Decade	Mineral
1		2 Krumm (A), Krum (BC)	Red jasper (hemag).
2		3 SeXep Krumm (A) Xer-Xepti-Krum (BC)	Chrystal next to gold
3	II Xont-hri		Emerald next to gold
4		1. Hā-tati (A), Hā-tat (BC)	Chrystal next to gold
5		2 Phu-tati (A), Phu-tat (BC)	Antimony next to gold
6		3 Tumti (A), Tumm (B) Tum (C)	Gold
7	III Sit-roh-pet (A) Sit-reh-en-pet (B)		Copper next to gold
8		1 Usit-bekt	Emerald
9		2 Ȧpisot (A) Ȧpesot (BC)	Iron next to gold
10		3 SobXot (A) SobXos (BC)	Crystal next to gold
11	IV Ā-pehuti reh pet (A) Mer-sit ā-pehuti-reh en pet ta (B)		Seer (stone) next to gold
12		1. tepā-Xont	Alabaster
13		2 Xont-hri	Ruby next to gold
14		3 Xont-Xri	Ruby next to gold
15	Vȧmseθ-ȧm (B)		Agate
16		1 θemat Xont	Crystal next to gold
17		2 Sopt-Xon	Ruby (next to gold)

Nº	Chief	Decade	Mineral
18		3 Her-ȧb-ua	Sapphire next to gold
19	VI Xu-neXeX		Chrystal next to gold
20		1 Sesme (A), Sešmu (C) θos-[ālk (B).	Chrystal next to gold
21		2 Konimme (A) Konime (BC)	Agate next to gold
22		3 Tpā-sal (A) Tpā-smal (BC)	Gold
23	VII Qa-mer-mulf (A)		Gold
24		1 Smali (A), Smal (BC)	Copper next to gold
25		2 Sråt (A) Sra (B) Sral (C)	Marble next to gold
26		3 Si-såt (A) si-sral (BC)	Agate next to gold
27	VIII Uet-ſem utá		Silver
28		1 SeXepti-sral (A) XerXepti [sral (BC	Silver (A) Crystal next to gold (B)
29		2 Tpā-Xu	Red jasper next to gold
30		3 Xu	Gold
31	IX Up-uat		Agate next to gold
32		1 Tpā-biu (AC) Tomm (B)	Gold
33		2. Biu (AC) Ušte (B)	Såma next to Gold
34		3 Xont-ḥri	Gold
35	X Ḥru-tep-nofir		Crystal next to gold (A), Gold (B)
36		1 Xont-Xri	Ruby on gold
37		2 Si-Kot	Crystal next to gold
38		3 Xou	Agate next to gold
39	XI Sam-neb-Xu		Crystal next to gold

No.	Chief	Decade	Mineral
40		1 ārt (AC), ārit (B)	Syenite next to gold
41		2 Remen-hri	Mennu next to Gold
42		3 Θos-ālk	Chrystal next to gold
43	XII Rā-m-hotp		Mennu next to Gold
44		[her ua (B)] 1 Uār (A) Uāret (C)	Gold
45		2 Tpā-sopdet	Ebony next to gold
46		3 Uśla-bikot	Ebony next to gold
47	The serpent Neker in the west over the world of life.		
48	Osiris (1. Leap day)		
49		Uśti (A), Uśle (B)	Ebony next to gold
50	Horus (2nd Leap day)		
51		Uśli (A), Uśle (B)	Ebony next to gold
52	Isis (4th Leap day)	[Var. B. Bikot	..... next to Gold]
53	1st	ānχ em χert (A) ānχ em śert-f (B)	Mennu next to Gold
54	Nephthys (5th Leap day)		
55		Senen	{Olive tree next to gold (A) Men next to gold (B)
56	The divine Sothis (1 Thoth)		Gold
57	I Anpru bast ta		Sapphire next to gold
58		1 Śela	Agate next to gold

Translator note: Brusch did not include the 3rd Leap day *(Set)* in his work.

End of decan list.

North Side, Strip B', Lower Display (Direction E-W)

The 12 hours of the night and the deities belonging to them

The individual hours are understood as female persons with a star on their heads. 

Next to each is a box in the Egyptian style with a closed latch door: 

Depending on the *number* of *stars* above, the *number* of the relevant *hour* is determined purely externally. The *eponymous deity*, distinguished by the scepter , stands next to the *hour goddess* belonging to it. On the following table the names of the hour gods are listed under A, the *names* of the eponymous *deities* under A', *the hour stars* under A'' and the *images* of the eponymous *deities* under A'''. Added to this list is 1. the list of hours of day and night on the *cubes of the facade* of the temple of Dendera (see plan pp. 2 and 3) under B, B' and 2. under C, C', C'', C''' the hour representations on the side of the *wall of the northern Osiris room* on the roof of the temple of Dendera. We note that the names of the *hours* in this epoch are borrowed from the *hour-gates* of the earlier epochs (see the Pharaohs table below), so that the ancient list given under C is the more correct one.

The 12 hours of the night and their eponymous gods[12]

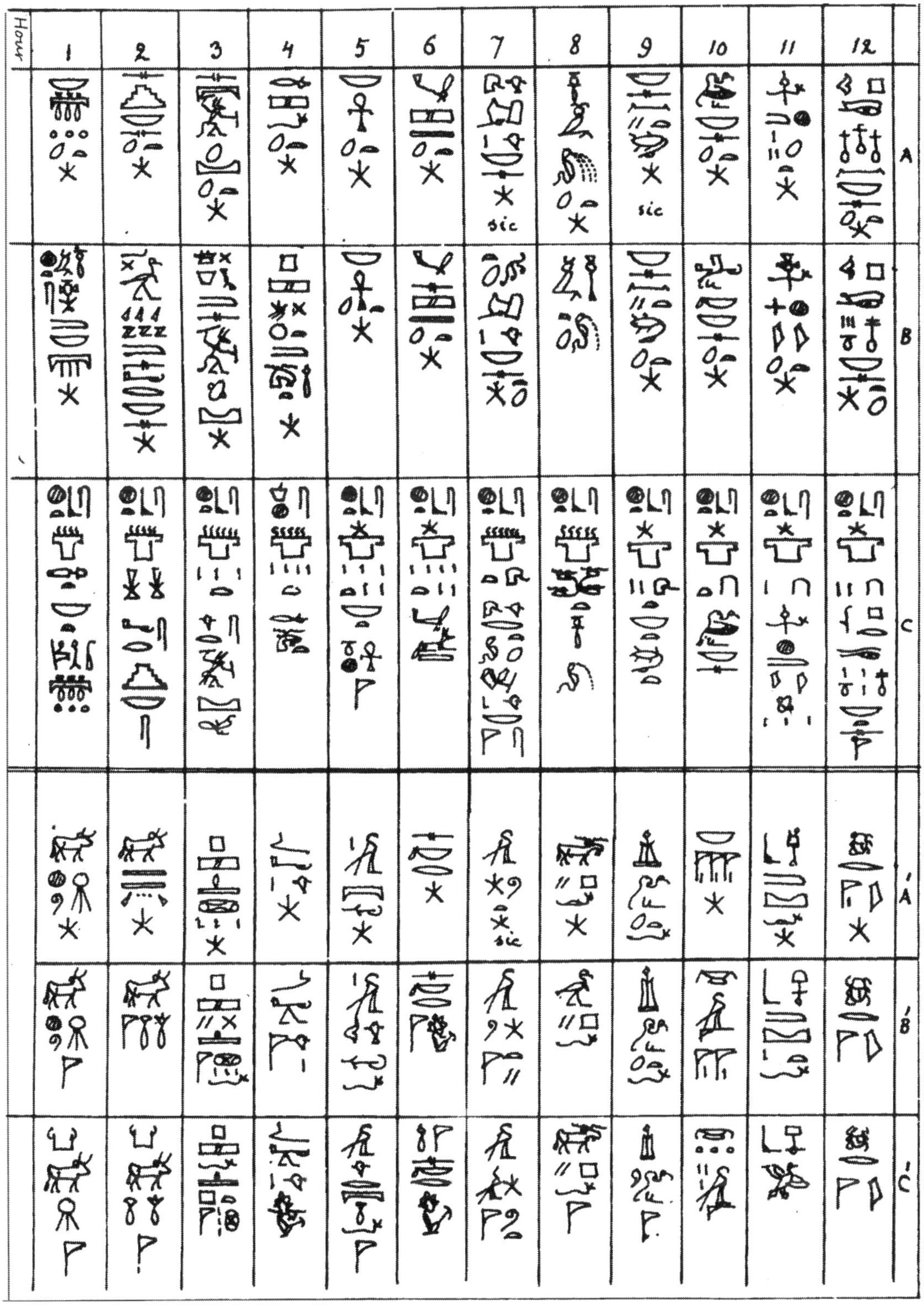

[12] Translator's note: Compare to the square zodiac in our book *The Secrets in the Secrets of Ancient Egypt.*

The 12 hours of the night and their eponymous gods (conclusion)

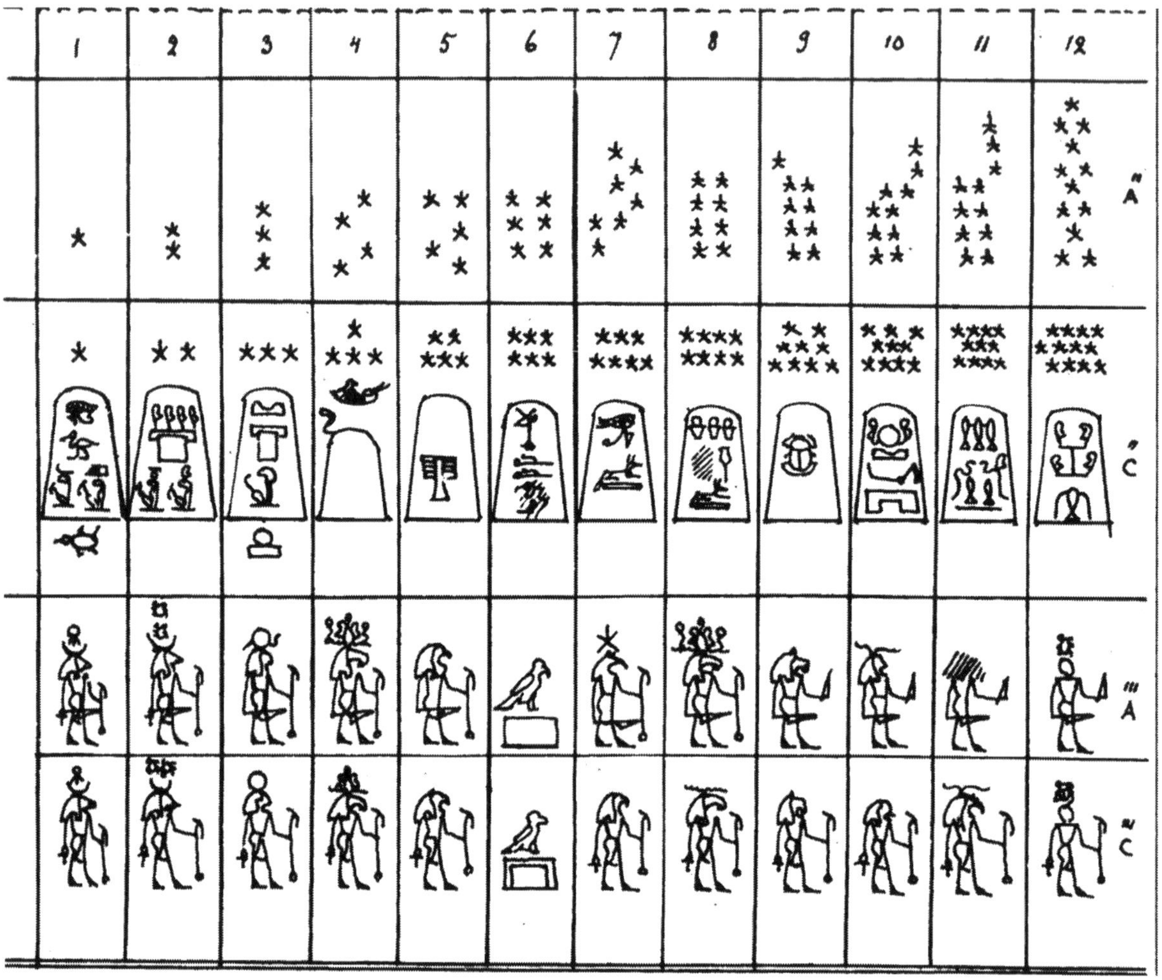

The last hour (i.e. the first after the actual sequence) is followed by three gods in this position: . The inscription next to the three men in adoring position describes them as 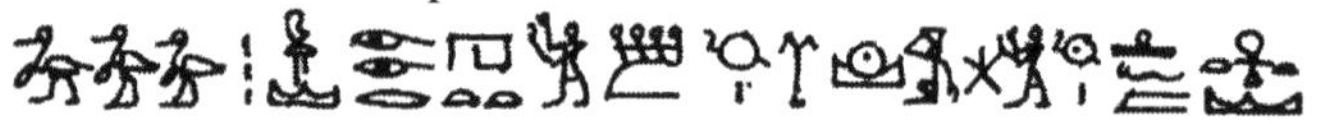 i.e. "the souls of the western region. The act of worship and the salutation to the sun god <u>Ra</u> is accomplished in his radiance, and the praise of the sun god <u>Ra</u> when he sets in the land of life" (i.e. in the West)

Three other male beings pull the sun ship with the image of the god (the evening sun) <u>*Atum*</u> on a long rope that ends in the front body of a Uraeus snake. They are referred to as:  "the stars (<u>*xem-sek*</u>) that accompany the sun in the northern sky." In another barge, which immediately follows the sun ship, the eye of the full moon shows itself in the form , to which the ibis-headed <u>*Thuti-Thoth*</u>  desires to express his homage. <u>*In front of the ships of the moon*</u> is the inscription:

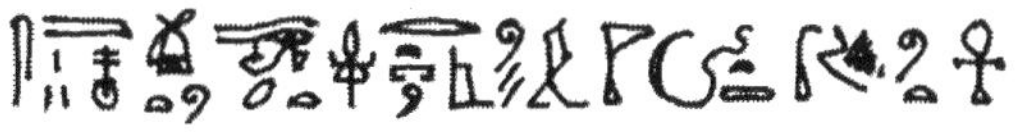 <u>*anxut mant tet iui er ustef aboti aperut emnafrus*</u>
"Life and renewal take place in eternity; the moon returns to its place and the orange full moon is endowed with its glory." A third ship shows the image of Osiris on the throne with crown  and scepter . In front of him, five stars in this order . A four-line text instructs about it:

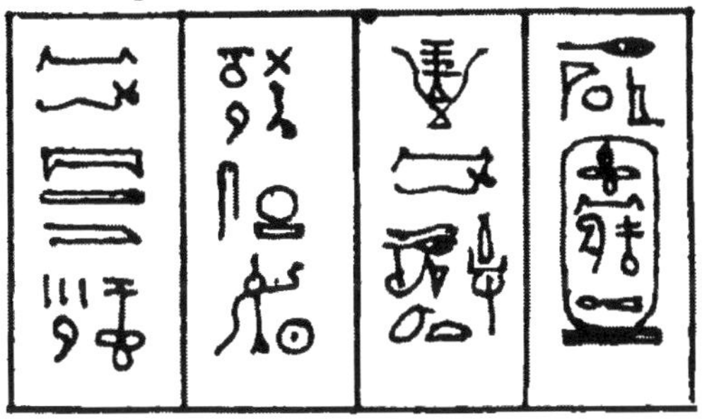

i.e. "Osiris-Onnophris, the triumphant, he has united with the full moon eye. He has repeated the cycle and he has illuminated heaven and earth with his glory." This is followed by three deities without (recognizable) inscriptions and, in reverse position, one of the four women standing up above the sky.

South Side, Strip B, Lower Image, Direction from E-W

The 12 hours of the day and their eponymous deities.

The individual hours of the day appear as a *goddess* with the sun disk  on her head . Next to her is the image of the eponymous deity belonging to her. For the sake of comparison, the corresponding part of the inscription B mentioned under the 12 hours of night has been added to the preceding list (A).

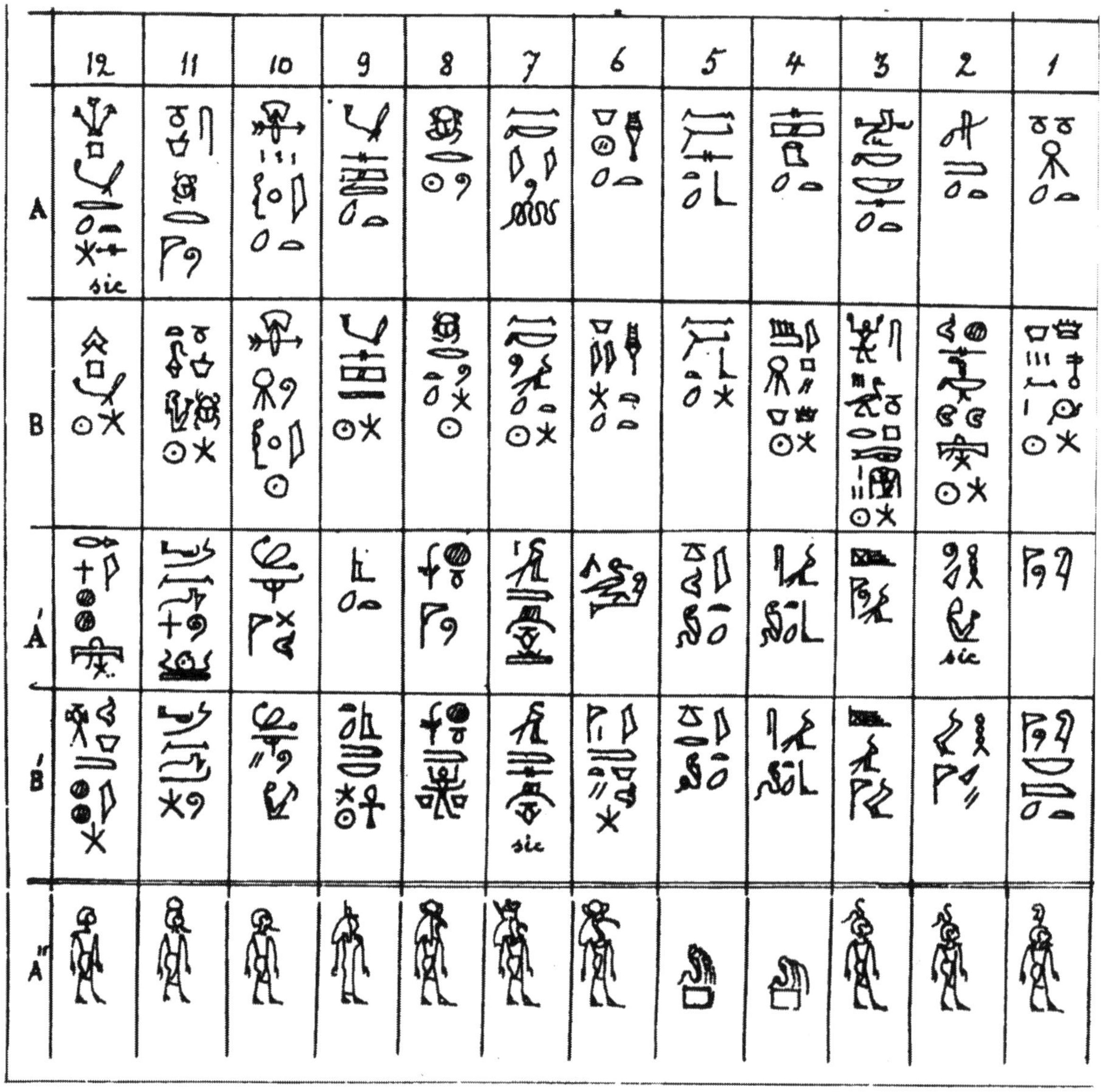

After this list of the hours of the day and their gods, 4 small ships follow. The first shows the image of a planet god  with the inscription 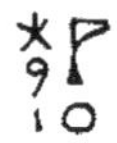 *nuter du* "god of the morning" i.e. Venus. In the second ship we can clearly see the figure of Osiris-Orion  under the name  In the third ship is the goddess Sothis: , and in the fourth a god with the moon disk on his head  is called:  i.e. aah "lunus." The following are attached to these representations. First of all, a large cargo ship, which is pulled by three men. The latter used the designation: 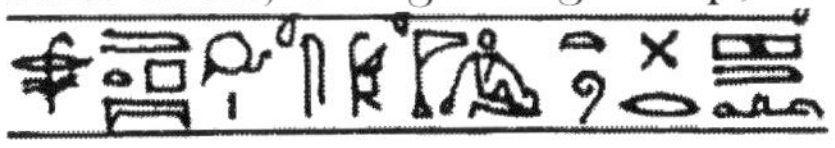 "the stars (sem-uerlu) which accompany the sun on the southern hill." In addition, the following short words: "Praise the sun every day." In the ship is the main character, the sun god called: ra hru-cuti "the sun, the shining Horus." Then three baboons[13] in this position . The inscription added to it says: "The Eastern souls. Praise to the god Xoper in his form, when he rises on the eastern side of light." 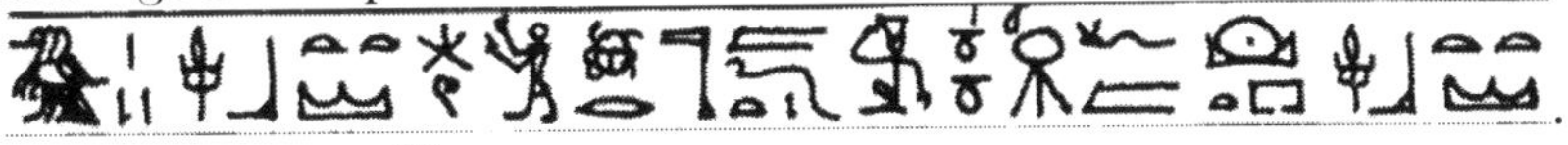. The god we are talking about now appears as  inside a naos, in the middle of a second magnificent ship. Above the god the inscription:  "Hail to you! You are changing into the shape of the god Xoper with the beautiful sun disc made of emerald." Three jackals pulling the ship bear the designation:

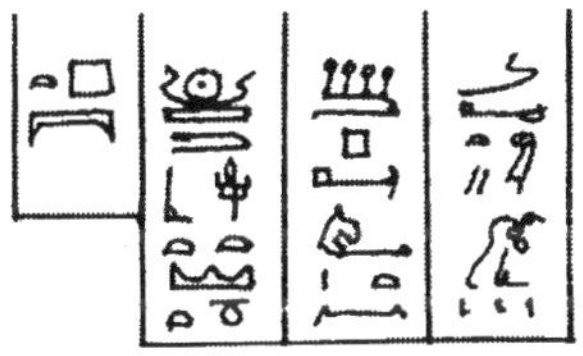

"the righteous grasp the cusp of the sun barque in the east of heaven."

[Note. In Esna, on the ceiling of the temple, there is a very similar image. The migrating jackals are named in the accompanying inscription:

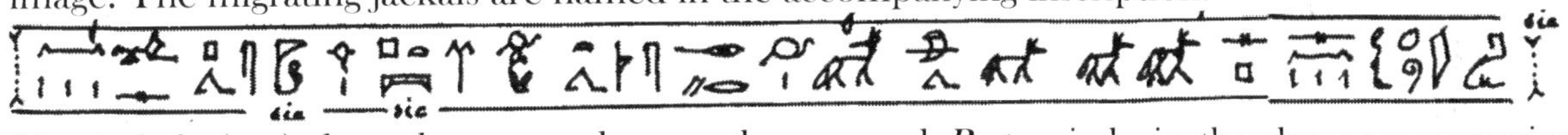

"the jackals (sap) draw the sun and cause the sun god Ra to circle in the sky, accompanying their Lord in their form."] The two deities behind them seated in a ship are no longer recognizable.

[13] The German word *Affen,* meaning monkey or ape, is translated here as baboon. See Translator Appendix for explanation.

North Side, Strip C' (from East to West)

The lunar sphere.

The overall representation is broken up into three separate sections, sequentially based on the waning, waxing and full moon.

1st Picture: The days of the waning moon.

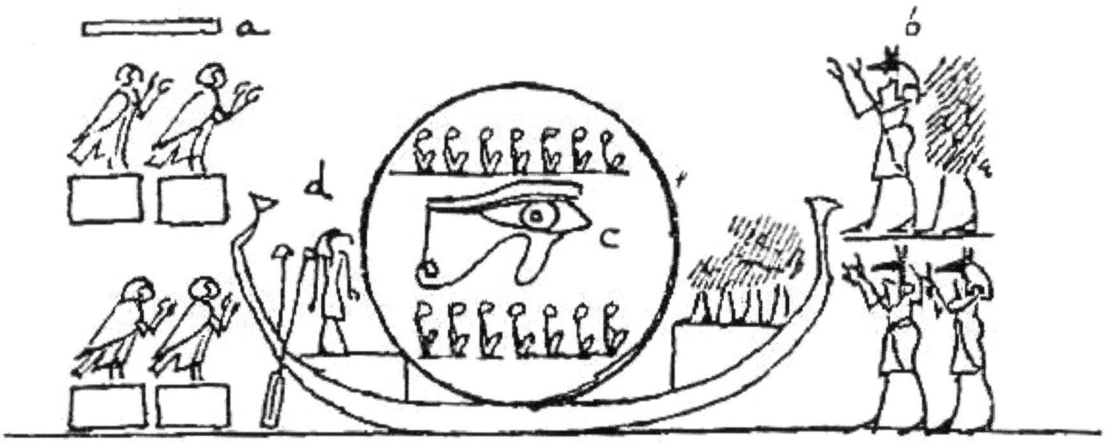

The 14 days of the Moon

In a boat, the moon eye *utat* is within a disc. Seven seated deities above and just as many below represent the 14 days of the waning moon. In connection with this, the current background is kept dark black. Four jackal-headed gods (b) worship the disk, as do four human-headed falcons (a) on the opposite side. A two-line inscription above the falcon reads:

i.e. "The spirits of the sixth day of the lunar month are the gods who glorify the moon's eye when it renews its cycle on the 15th. Behold the god in his form as a splendid child he has fashioned from the lunar eye with its (the eye's) splendor." The god is *Thoth*, the same one who is in the boat at (d).

Towards the west, behind the image just described, there is a goddess and the god *Thoth*, both in a worshipful position, paying homage to five  standing above each other. The writings on it are illegible.

2nd Picture. The 14 days of the waxing moon.

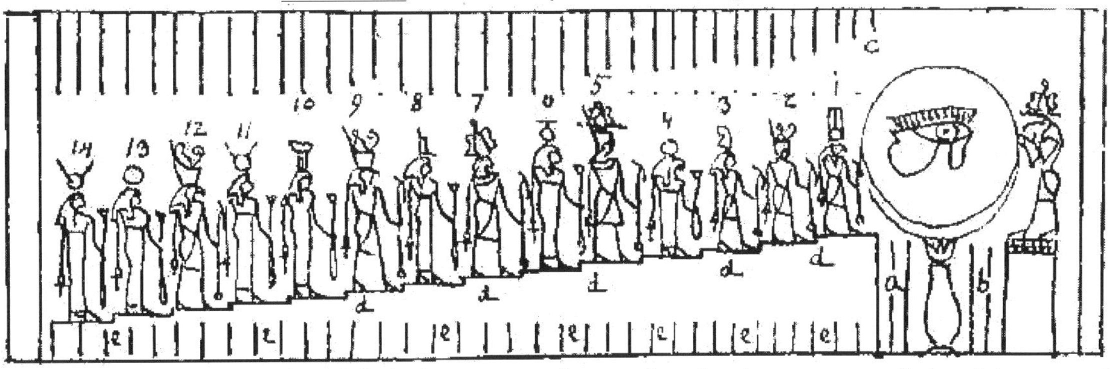

Fourteen deities, each of which belongs to a lunar day, begin on one of the 14 steps of a staircase (1-14). At the upper end of it, the full moon shines towards them, floating on a pillar. Behind, the god *Thoth* with hands raised in adoration.

Inscriptions accompanying the above illustration.

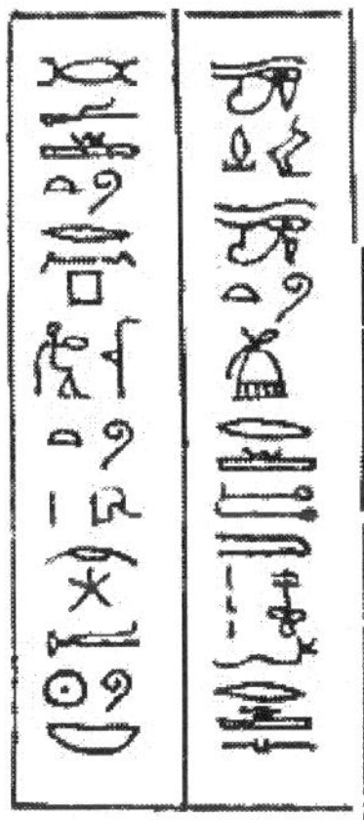

1) a. to the left of the column with the moon-eye at the top, downwards: "the moon-eye (the full moon) is intact and it is endowed with its glories for blessing, it is protected and it rejuvenates every month."

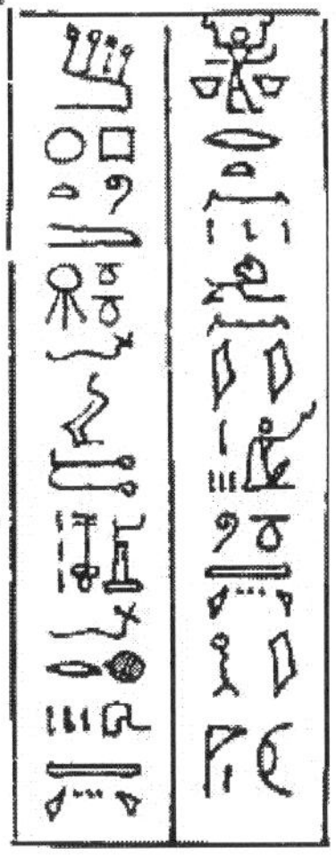

2) b. To the right of the column downward: rejoice, you inhabitants of the earth! The moon shines when it rises and its ship, seat of his glory, is intended for those who dwell on earth."

3) c. over 14 deities on the steps of the lunar staircase, a longer text to which we have added the variants of an identical inscription that can be found on the lunar staircase on the roof of the temple of Dendera. The corresponding passages are indicated in the copy on page 32 followed by vertical lines under the letter f.

The 14 Days of the Waxing Moon

The fortnight of the waxing moon

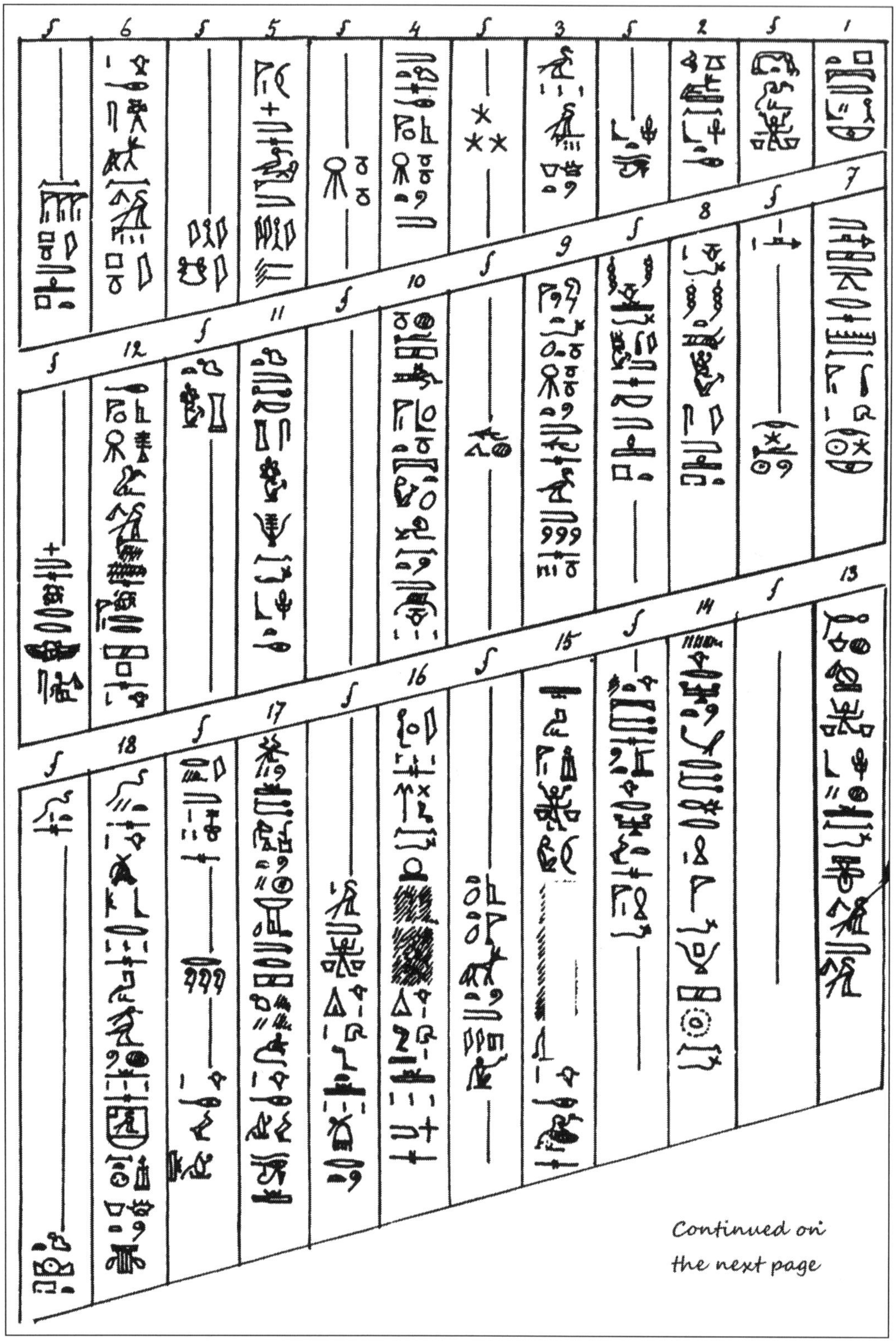

Continued on the next page

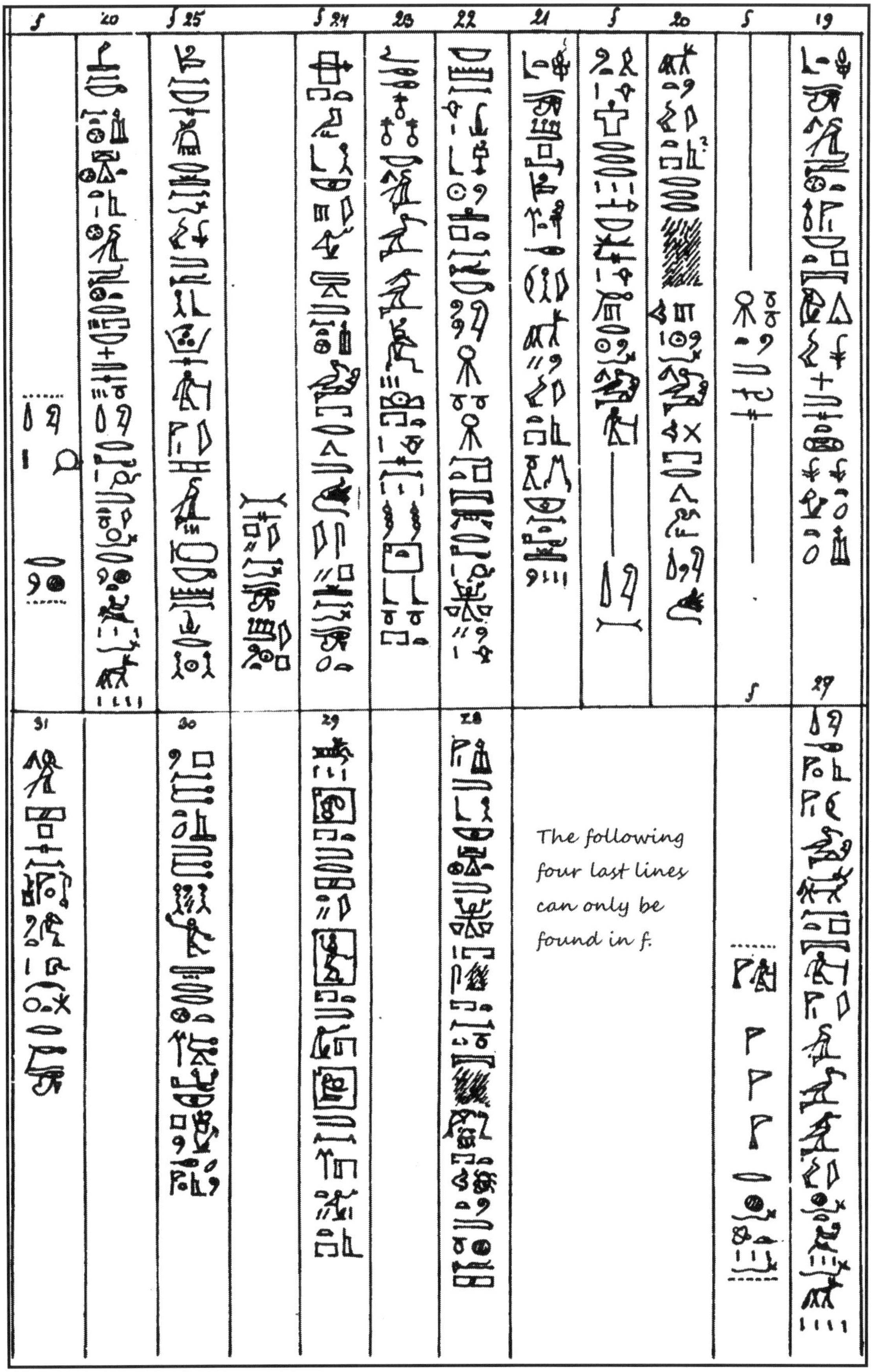
f
20
f 25
f 24
23
22
21
f
20
f
19
31
30
29
28
The following four last lines can only be found in f.
f
27

Translation of the previous Text.

"The heavens are joyful [and the var. in f: the firmament is full of joy], in which they have the form of the full moon. The souls of the gods appear in him and Osiris rises up in him as moon god. *Thoth* shows himself as a protector to watch over it. The individual deities approach, as they stride towards it.
The moon god (1st lunar day) every month his heart is filled with delight.
The god Atum as the (2nd lunar day) is satisfied.
The god Su (3rd lunar day) and *the goddess Tafnut* (4th lunar day) rise up in him and the soul in her body is delighted.
The god Qeb (5th lunar day) and *the goddess Nut* (6 lunar day) are happy when the god *Xont-makes* (i.e. Osiris) has united with the moon eye.
The god Osiris (7th lunar day) shines as God in him, a splendid beetle, he fills in what is taken away and is full of joy that he has successfully united god with god. The height of heaven rises up brilliantly [f: heaven is raised up and ascends] while carrying his majesty. He lights up the earth as the god *An*. The moon god rejoices.
The divine Isis (8th lunar day) night full of merriment to grant protection to his form as he renews his cycle.
The god Horus (9th lunar day) is joyful and gives the respective filling with his glories [column 20, Dendera: for repeated renewal and rejuvenation].
The goddess Nephthys (10th lunar day), in delight, shields his figure and fills his parts with his glories.
The goddess Hathor (11th lunar day), the Tentyrite, appears in the Moon's Eye.
The god Horus (12th lunar day). Apollinopolitis, the great God and Lord of Heaven, shows himself in him [f: is absorbed in him]
The goddess Tanent (the 13th) and *the goddess Anet* (the 14th lunar day) come to the place [? f: come later]. Every part of him fills his day.
Thoth (15th lunar day) emerges victorious. The moon eye is greeted by the sun eye. The moon comes to the (right) place without introducing a fixed date, (because) all its laws are regulated for rising and setting. You are the shining ray for heaven and earth and the sun god Ra is pleased to behold your glory. The gods of the sphere of light, their lord is full of bliss. The *Hatbenben* temple of the great sanctuary (in *Ôn*) is in a festive mood. Merriment reigns all around *Ant* (Tentyra) when *Thoth* appears as the victor, when he has prepared the full moon for its owner, when he has filled it with what it requires. Osiris, friend of the gods, your name is permanent! You have taken possession of the city of *Ant-Adut* (Tentyra), of the Horus city of Apollinopolis and all the temples in it. The sun god Ra triumphs in his disc over his opponents - 4 times - Osiris, the moon god *Thoth*, the bull of heaven [or: the lord of heaven], the prince of gods, triumphs over his enemies - 4 times."

In the four final lines (after f), the thought is repeated how the city of Tentyra (called by nine different names) gives itself to joy when "the glorious soul of the god Osiris rejuvenates every month to take possession of the full moon, the lunar eye *utat*."

At the foot of the gods mentioned, under the individual steps of the lunar staircase there is a repetition of their names (under d) with the regularly recurring addition  ...

 filling the eye (of the moon) with its .... "The places indicated by dots contain the first instructive synonym for the idea <u>*eye*</u>," the latter contain various expressions that relate to the object that fills the eye. Among the d' and d", variants that I borrowed from the terrace walls of the Temple of Dendera. D" mentions stone and plant names associated with the moon's eye.

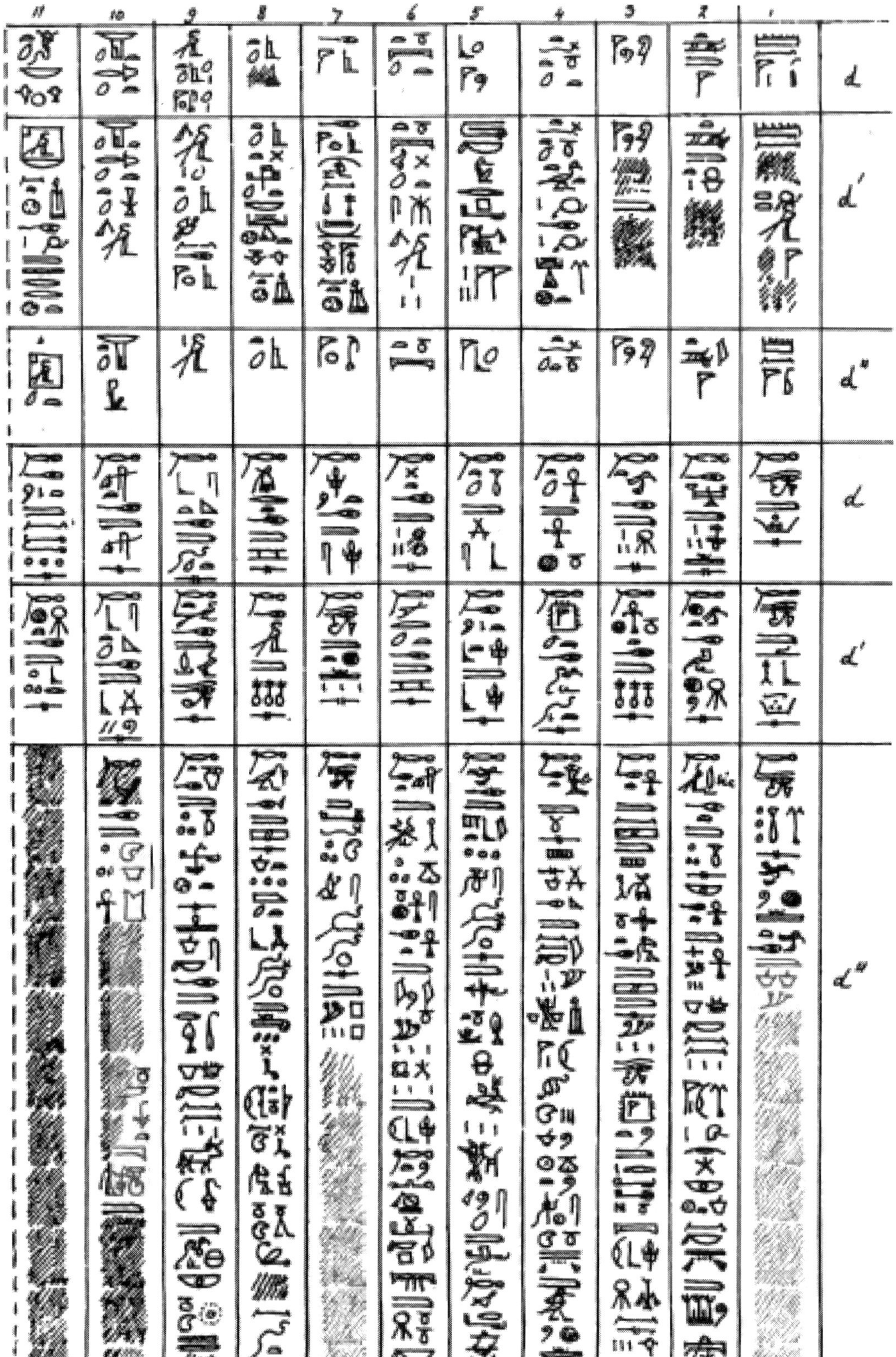

Leteon. Insche.

Translation

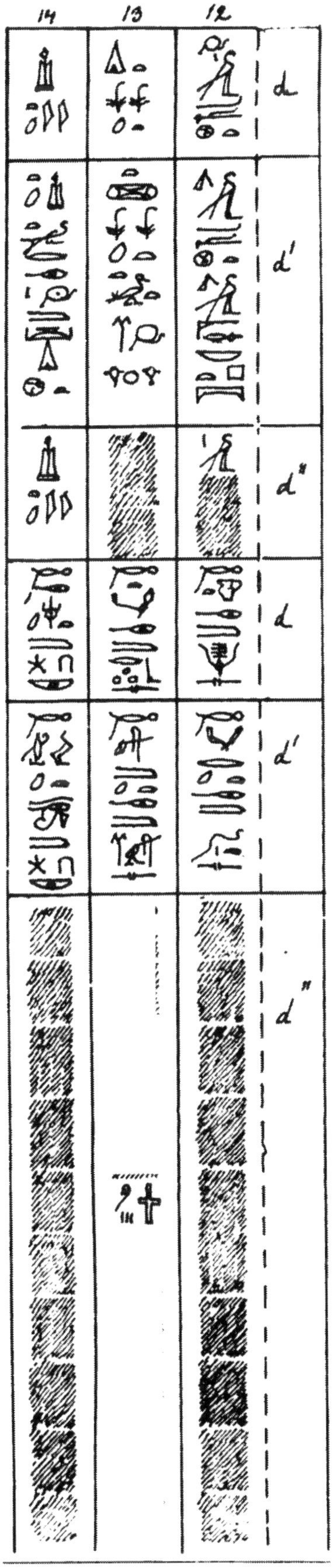

1. *God moon* [d' *Ra-Hru-Xuti,* the great god] the full moon eye is filled with its requisite (d, d') [the full moon eye is filled with green stone, the luminous eye is made shining with the plant *beb....* d"].
2. *God Atum* (d, d") [God Atum, the Father.....d']. The Horus eye is filled with its best (d) [the shining eye is filled with its light, d'- the full moon eye is filled with emerald, the eye of life is festively decorated with the tree of life. You appear to us as the moon every month, you illuminate the earth in the evening." [d".]
3. *God Su.* The shining eye is filled with its light (d) [the eye of life is filled with its best d'_ the eye of life is filled with ruby, the youthful eye is decked out with ears of corn. The sanctified eye is on our hand. The moon he illuminates the faces of men, d"].
4. *Goddess Tafnut* (d, d") [*Goddess Tafnut,* the daughter of Ra, the eye of Ra in Tentyra d']. The eye of life is filled with life (d) [the holy eye is filled with its form d'- the rimmed eye (*mert*) is filled with alabaster, the anointed eye is provided with the Annek plant. O moon, when your rays shine, you illuminate the earth with your light!
5. God Qebu [God Keb, the splendid heir to the throne among the gods d'] The prosperous eye is filled with what it covers (d) [the left eye is filled with what it demands d' - the shining eye is filled with marble, Its pupil is given prosperity by the wood of the *kal* tree (what follows is incomprehensible to me d"].
6. *Goddess Nut* (d, d") [*the goddess Nut,* the great one, the mother of the gods d']. The big eye is filled with its .... (d)[the peripheral eye is filled in with what it loves d'_ the leading eye is filled with *hemag* stone (red jasper?), the life eye gains life through the *Iun* plant. The matter in the moon is fully connected and nothing is missing. The clouds are scattered at your rising, d"]
7. *God Osiris* (d, d") [*Osiris onnophris,* the triumphant, the great god in Tentyra, d']. The left eye is filled with what it longs for (d) [the full moon eye is filled with all its things d' - the full moon eye is filled with emerald, the pupil is protected by the pepe plant .... d".
8. *Goddess Isis* [*Isis,* the great mother of god, mistress of Tentyra, in the (local) temple of Anet d']. The stately eye is filled with what it likes (d) [The Horus eye is filled with its perfections d'- the red eye is filled with sapphire, the pupil is covered with barley. O you with a returning figure, you renew yourself without finding an end, in eternity, d"].
9. God Horus, son of Isis, son of Osiris (d) [Horus, son of Isis, heir of Osiris d', Horus d"]. The whole eye is filled with its

physical components (d) [the big eye is filled with that which brings it salvation d'- the friendly eye is filled with Upper Egyptian green stone and the anointed eye is satisfied with the willow tree. You rise up, oh you fiery bull, on the day of the new moon. You shine with .... d''].

10. *Goddess Nephthys* the benefactress (d) [*Nephthys*, the benefactress, the sister of *Horus* (d') - *Nephthys* d'']. The leading eye is filled with what it guides (d) - [the blessed eye is filled with what it covers a'- the great eye is filled with emerald .... d''].
11. *Hathor*, the mistress of Tentyra d [*Hathor*, the mistress of Anet, the eye of *Ra* in Tentyra d'-*Hathor* d''] - The eye is filled with its substance (d). [The shining eye is filled with its pupil d '] (text d'' destroyed).
12. *God Horus* of Apollinopolis (d) [*Horus* of Apollinopolis the great god, the Lord of heaven d'- *Horus* of....d'']. The kind eye is filled with what is friend to him (d) [The great eye is filled with its physical components d'].
13. *Goddess Tanenet* (d) [Goddess *Tanenet*, the daughter of *Ra* in Tentyra d'] The static eye is filled with its own pupil (d) - the false eye is filled with what it directs d'].
14. *Goddess Anit* (d) [*Anit*, the great one, the eye of *Ra* in Tentyra d'-Anit d'']. The left eye is filled out on the 15th day of the lunar month (d) [the full moon eye is filled out on the 15th day of the lunar month d' -- text in d'' destroyed, as in Col. 11, 12, 13.

e. List of the Thirty Days of the Lunar Month

according to their eponymous names and list of the protective deities belonging to them.

The same is located just below the lunar staircase, which is described above. For the sake of completeness and for the sake of comparison, the same list (from Ptolemaic times, i.e. older than that of Dendera) is added under e' according to the astronomical calendar picture on the upper edge of the north wall of the pronaos of Edfu, further under e'' the list of the first 19 lunar days according to a text in the second room of the northern Osiris temple on the roof of the great temple of Dendera, and finally under e''' occasional variants from the old (A), new (N) and Ptolemaic empires [P].

Heading: 

"Names of the 30 Moons"

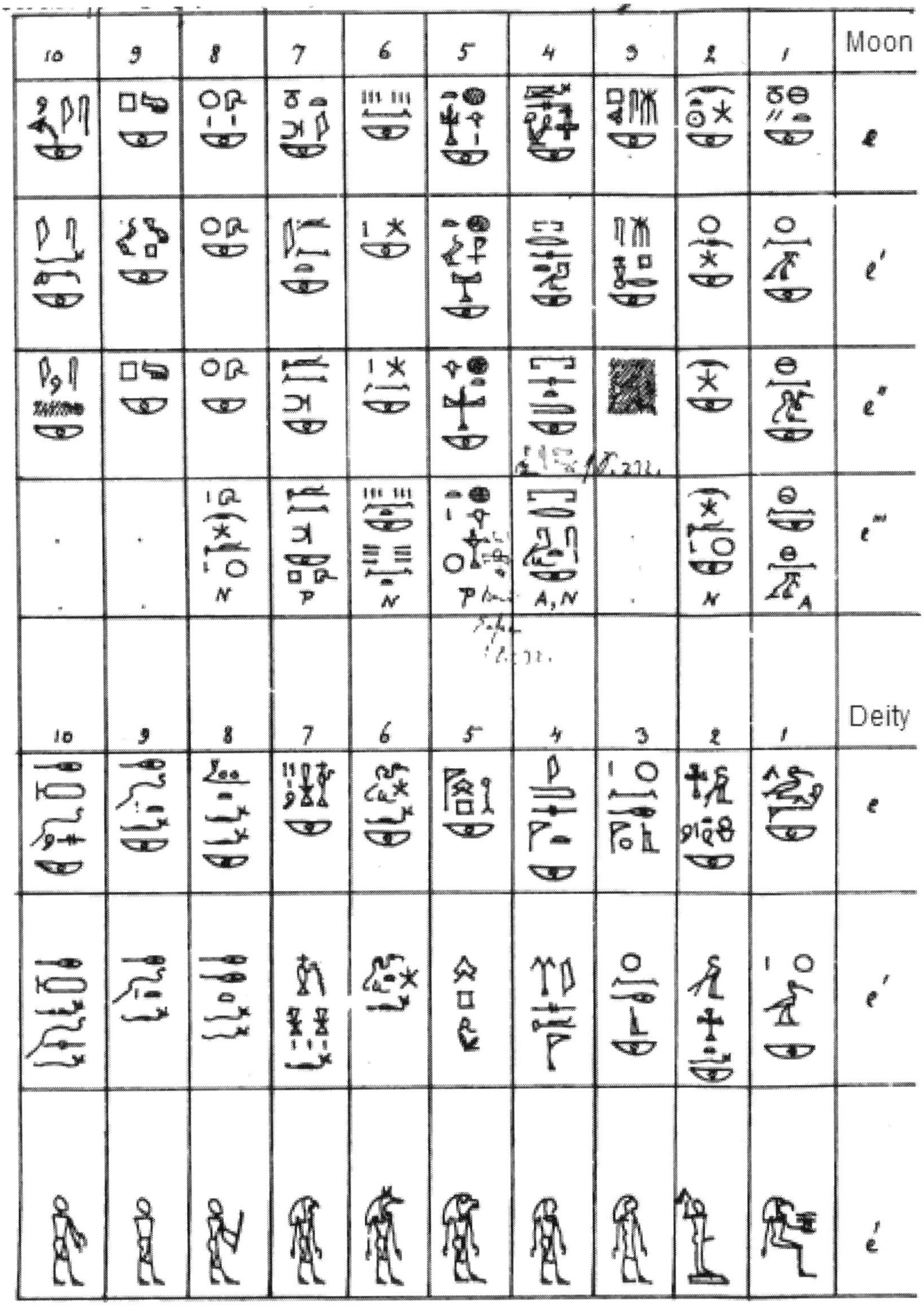

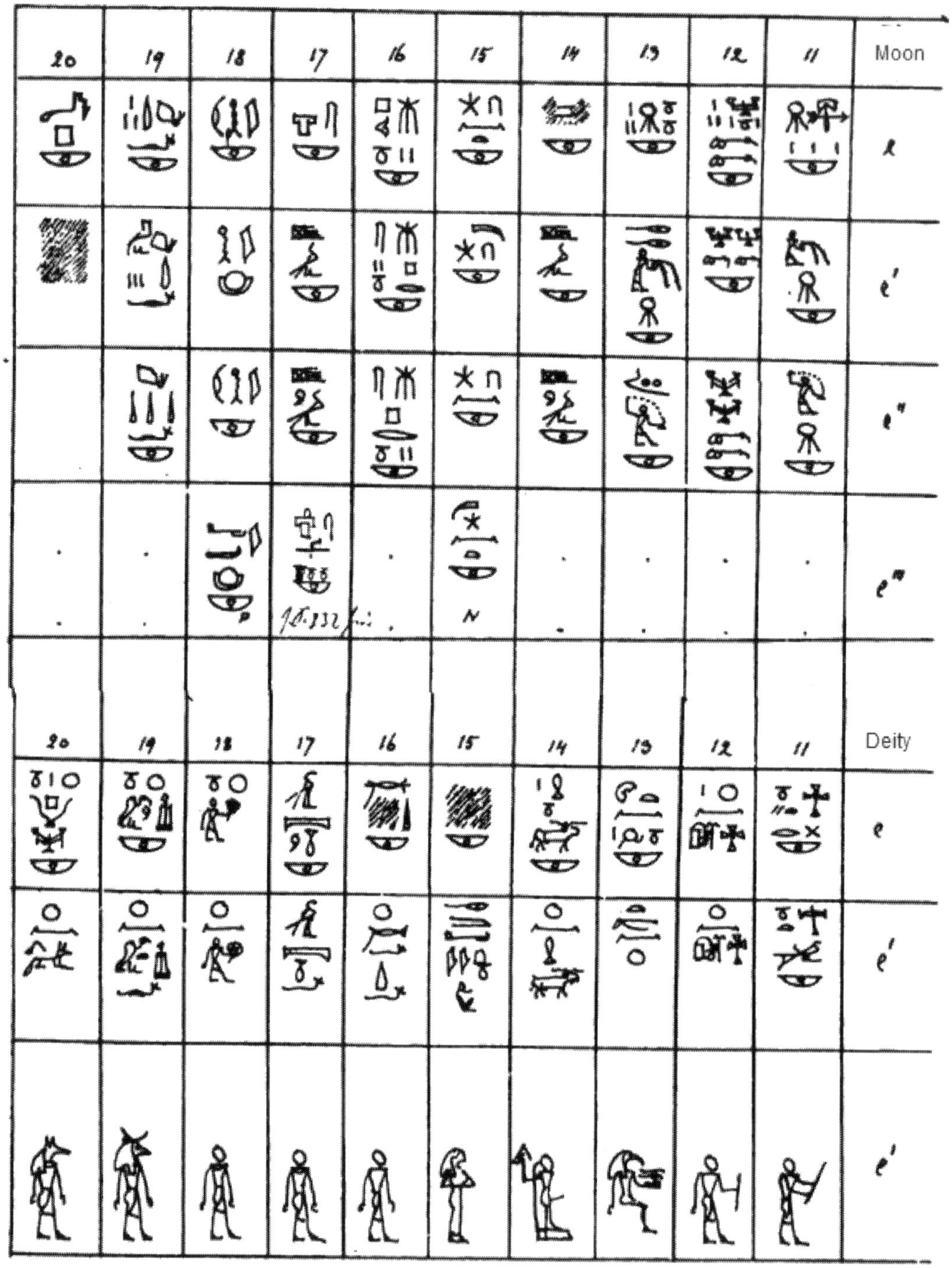

20 19 18 17 16 15 14 13 12 11
Moon
e
e'
e''
e'''
Deity

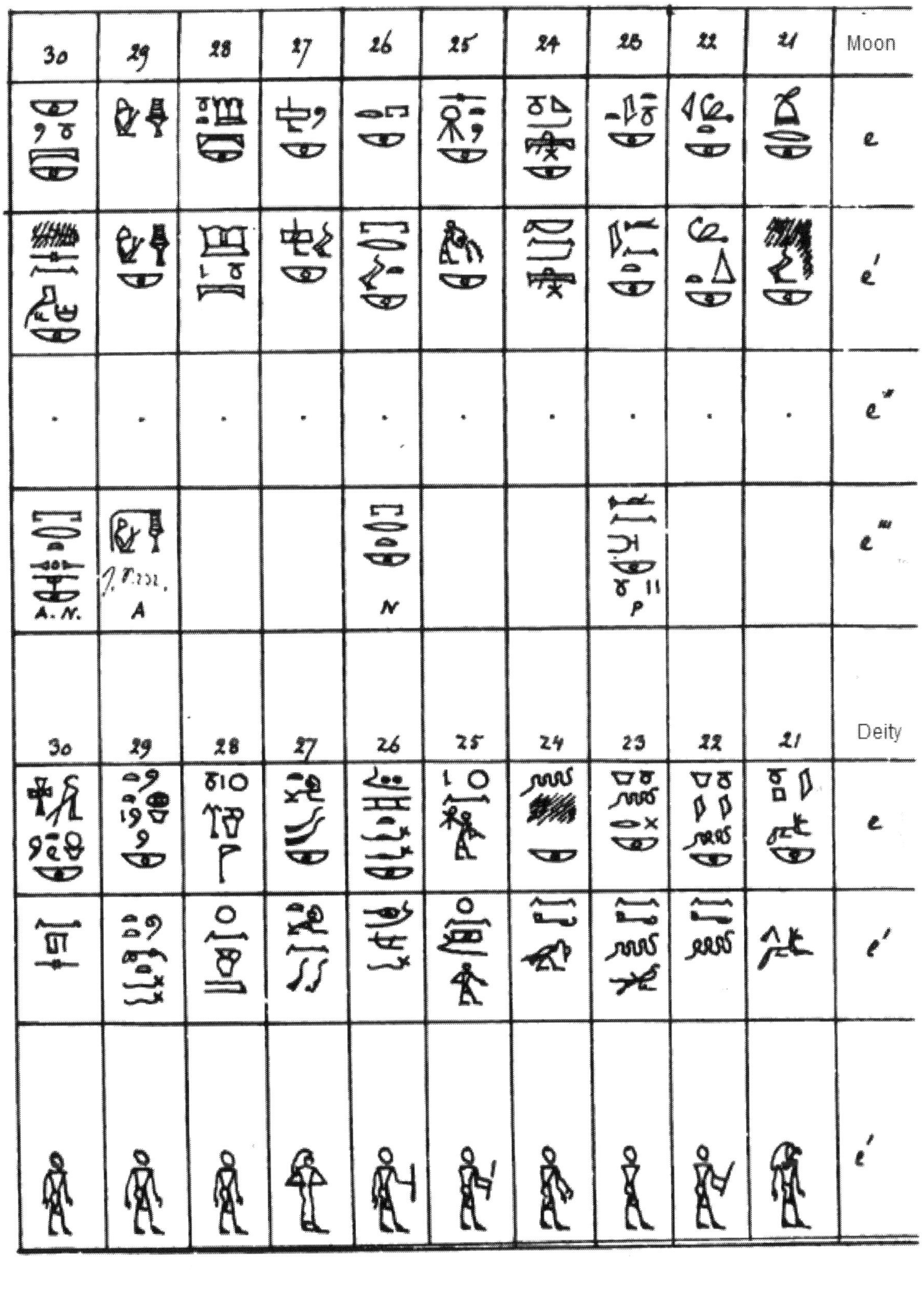
30 29 28 27 26 25 24 23 22 21 Moon
e
e'
e''
e'''
A. N.
A
N
P
30 29 28 27 26 25 24 23 22 21 Deity
e
e'
e'

Translation

1st lunar day. *Hib-enti-paut* "celebration of the new moon", also after the deity consecrated to him *hib-Thuti* "festival of *Thoth*" or *haru en Thuti* "day of Thoth" (e-c).

2nd lunar day. *Hib-abud* (*abut*) "celebration of the month", at the same time *hib Hru-at-f* "celebration of the god Horus, the avenger of his father."

3rd lunar day. *Hib-masper* "*Masper* celebration" (e) or *hib-masper-tep* "celebration of the first *Masper*" (see below 16th lunar day), at the same time *haru-en-Usari* "Day of Osiris."

4th lunar day. *Hib-pir-setem* "celebration of the apparition of the *Setem*, "at the same time *hib-amset* "celebration of the god *Amset*."

5th lunar day. *Hib-Xet-hr-Xau* "celebration of the sacrifice on the altar" at the same time *hib-Hapu* "celebration of the god *Hapu*".

6th lunar day. *Hib-en-(*var. *ent)-sas* "celebration of the Six", at the same time *hib Duamutf* "celebration of the god *Duamutf*."

7th lunar day. *Hib tena* (*dena, den*) "celebration of the section," var. *Hib den tep* "celebration of the first section," at the same time *hib-gebh-sonuf* "celebration of god *Gebhsonuf*."

8th lunar day. *Hib tep-sop*, "Festival of the beginning of the *Sop* (?)", var. (e'") *tep-abud* "the beginning of the month," at the same time *hib-ma-at-f* "Festival of the one who looks at his father" (name of a deity).

9th lunar day. *Hib-kapu* "celebration of concealment", at the same time *hib-ar-tet-f* "celebration of the one who created his body" (name of a deity).

10th lunar day. *Hib-sat* "celebration of purification", at the same time *hib-ar-rant-tesef* "celebration of the (god) who created his name himself."

11th lunar day. *Hib-situ* "celebration of the expulsion of light", at the same time *hib net nuti uer* "celebration of the great work".

12th lunar day. *Hib-herher* "celebration of?" at the same time *haru en net* "day of grinding."

13th lunar day. *Hib-uben* "celebration of the shining dawn" (e) or instead of *hib-uben* in e' and e" *hib-ma-sit* "celebration of the sight of the projection of light," at the same time *hib-teken en ra* "celebration of the approach of the sun."

14th lunar day. *Hib-sa* (or *sau*) "celebration of knowledge," at the same time *hib hon en ba* "celebration of the majesty of the ram" or *Haru en hon ba* "day of the majesty of the ram."

15th lunar day. *Hib-en (ent)-met-dua* "the celebration of the fifteenth" (more precisely: "the fifteen"), at the same time *hib ar-maui* "celebration of the god *Armaui*."

16th lunar day *Hib-masper-Son-nu* "celebration of the second *Masper*" (compare above, the 3rd lunar day), at the same time *hib* (or *haru*)*-sad-gemed-f* "celebration (or day) of the (god) who utters his speech."

17th lunar day. *Hib-sa* (*sau*) "celebration of the recognition," at the same time *hib-hru-hr-uot-f* "celebration of *Horus* on a pillar."

18th lunar day. *Hib-ah* (var. *Aah*) "celebration of the moon", at the same time *haru en ahi* "day of boys."

19th lunar day. *Hib-solem-gemeduf* "celebration of those who hear his speeches," at the same time *haru en An-mutef* "Day of God's grace."

20th lunar day. *Hib-anp* "Festival of *anp*," at the same time *haru en up-uat* "day of opening the way" (special name for *Anubis*, which, for example, has been clearly inherited from the Greek paraphrase of the Egyptian personal name *Pet-up-ua*.)

21st lunar day. *Hib-aper* "celebration of furnishings," at the same time *hib anup*, "celebration as *Anubis*."

22nd lunar day. *Hib-phu-sopdet* "celebration of the end of the triangle," at the same time *hib nai* (or) *na*) "celebration of the snake *Nai.*"
23rd lunar day. *Hib tenat* (var. *denat*) "celebration of the section" or *den-sonnu* "of the second section" (compare above, the 7th lunar day), at the same time *hib na uer* "celebration of the great serpent *na*".
24th lunar day. *Hib qenh* (var. *kenh*) "celebration of darkness" at the same time *hib-na-dost* "celebration of the red snake *na.*"
25th lunar day. *Hib situ* "celebration of the throwing out of light" (*hib situ* already mentioned the second lunar day), at the same time *haru en senn* "day of the deaf".
26th lunar day. *Hib-pir* (*pirt, pirut*) "celebration of the apparition", at the same time *hib ma-mi-atef* "celebration of the (god) who sees the friend of his father."
27th lunar day. *Hib-usb* "celebration of *Useb*," at the same time *hib do abui* "suitor of (god) with a raised pair of horns."
28th lunar day. *Hib-sed-ent-pet* "celebration of the tail festival of heaven", at the same time *haru en Xnum* "day of god *Xnum*".
29th lunar day. *Hib-aha-ar* "celebration *Aha-ar*", at the same time *hib-utet-at-f* "celebration of the Archangel of his father."
30th lunar day. *Hib-nu-pet* "celebration of heaven" (e) *hib... s-ne hem* (e'), in A. and N. *hib pirt Xim* "celebration of the apparition of god *Xim*," at the same time after e: *hib Hru net atef* "celebration of Horus the avenger."

Note. The earliest traces of this voluminous list of lunar days and the deities belonging to them go back to the times of the XVIII and XIX dynasties. I refer primarily to the astronomical ceiling paintings in the tomb of King Seti I and in the so-called Ramesseum at Thebes from the time of Ramses II. At the places mentioned, to the right and left of the constellations located near the North Pole, there is a kingdom of divine persons whose names and forms (the latter always with a disc on their heads) correspond to the names and personifications of the registers of Dendera and Edfu. The deviations are only to be settled owing to the more recent lists. On p. 44 I give the (incomplete) list from Seti's tomb and from the Ramesseum and leave it to the reader to compare it with the names and figures given above. The agreement is unmistakable. Otherwise the same names and positions appear again, mostly after a certain selection, in individual inscriptions, evidently to point to the individual phases of the moon. As an example I cite the inscription I gave in *Zeitschrift*, 1881, p. 93, from the large Osiris text of the Temple of Dendera.

Lunar Days from the Era of the XIX Dynasty

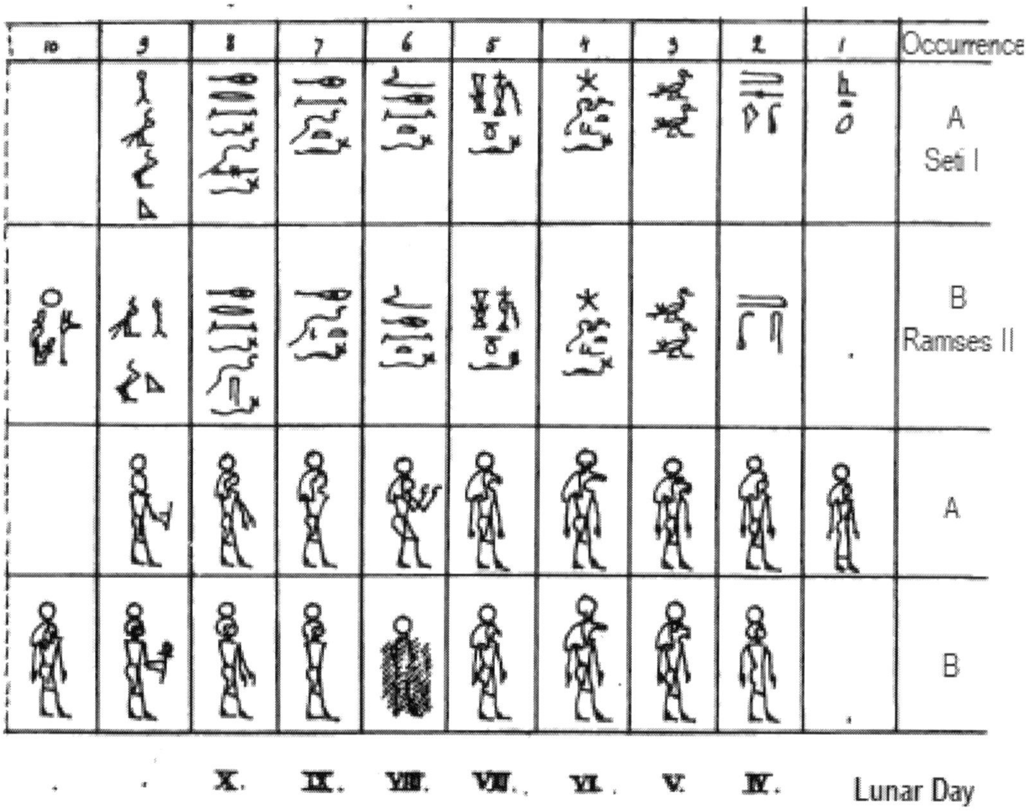

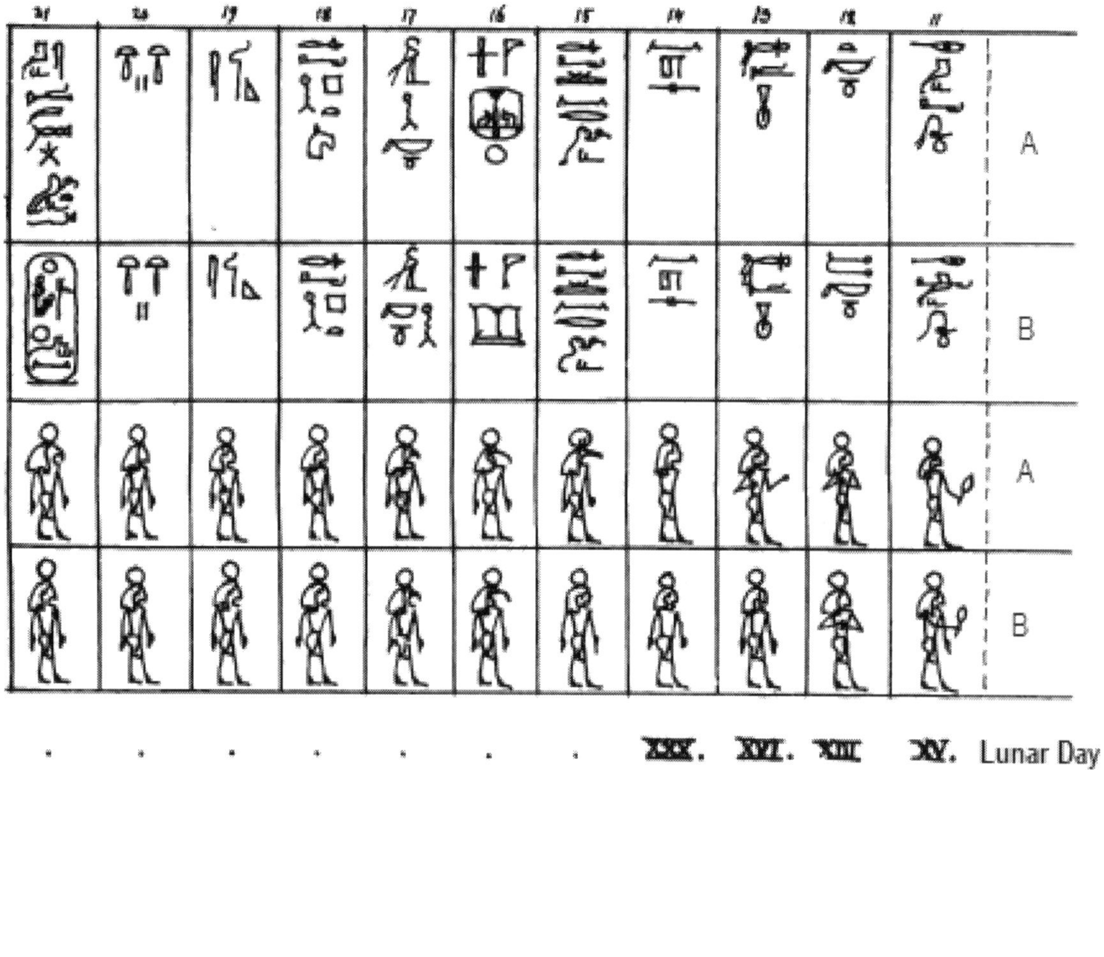

At the end of the important list of lunar days from Dendera there is the not unusual formula recorded here on the right: "All life is with them, all rest is with them, all power is with them," which concludes the inscribed part of the two pictures.

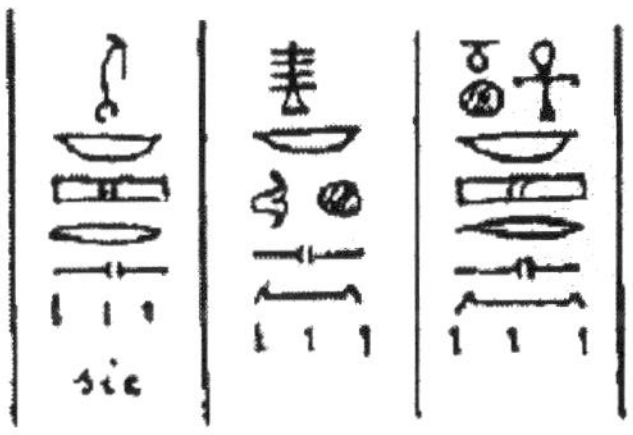

III. Image. Apotheosis of Osiris as a moon god.

The god is enthroned in a ship, Isis in front of him, and Nephthys sitting behind him; the ship rests on the sign for the sky , which is carried by four goddesses, the latter in turn designated as east, south, north, west." Above Osiris the inscription: "Osiris Onnophris, the triumphant, has entered the lunar month," in other words on the full moon.

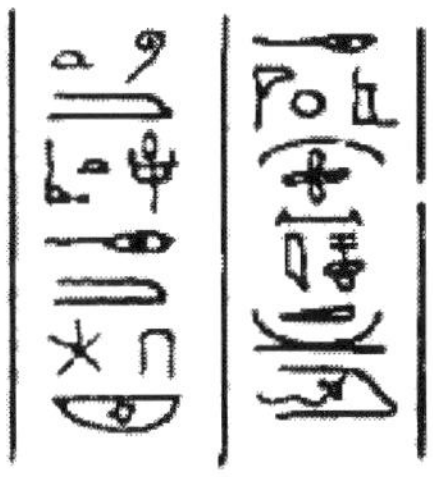

About *Nephthys*, the text: 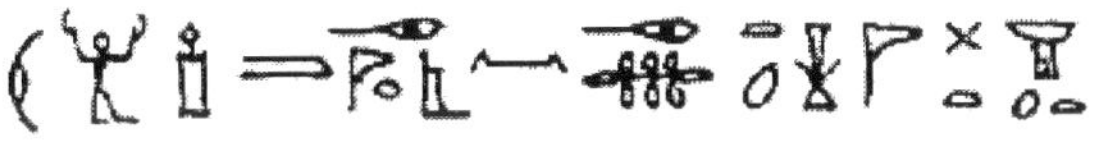 i.e. "*Nephthys*, the great, the sister of god, gives protection to *Osiris* as the moon." To the right of the ship the spirits of *Buto* (symbolizing the northern region), to the left of it, with jackal heads like those with sparrowhawk heads, the spirits of *Xen* or *Nexen*, Eileithyiaopolis. Behind this whole idea, to the west, the four pairs of the so-called "eight," the men with frog heads, the women with snake heads. Behind it, in the west corner, the wind animals,  i.e. the *North* and  the *West*, as in the east corner opposite:  [*the East*] and  *the South*.

South Side Strip C Middle Illustration (E-W)

Twelve ships, each with a sun disk in the middle, each disk with a picture on the inside, serve to express the sun moving from east to west in the twelve hours of the day. There is great confusion in the order and drawing of these in the publication of the *Description de l'Egypte*. In the narrow long columns under the sun ships, the following text:

(Difficult to read at first, it contains the names of the children) "[The sun god *Ra*] he rises in the morning, a wonderful boy nourished by a cow, an infant with limbs strengthened for the day. A child in the morning, a boy at lunchtime, he is *Atum* in the evening. Arrived in the west, its traditions take their course: the *Maad*-barge is located in its place from yesterday. The emerging light becomes an incarnate one in the womb and rises in the east every morning. Crossing the sky there is no rest for him. If he has circled the heavens with a happy heart, his heart rests in his solar body and in his summer body. He is the king of time.

Another narrower text over the sun ships, also beginning with the names and titles of the daytime sun, reads as follows: "If he shows himself in the place of yesterday, the gods and goddesses praise him at his rising and people pray to him every day. The eyes do their work, the ears open, the nose opens, the windpipe breathes, the Lord of the Holy Lord guides and directs what fills the heart, the Lord of Protection preserves the limbs [what follows is incomprehensible to me, afterwards:], the air comes out of his nose to give life today and forever."

On the next table there is a clear representation of the 12 sun disks in their sequence with accompanying inscriptions (under A, a). The pictures and inscriptions dealing with the same subject from the pronaos of the temple of Edfu are attached as instructive variants (under B, b) and the corresponding representations on the wooden coffin of a certain *Xaf* in the Museum at Bulag (compare the following under the title: The chapters of the metamorphoses, etc.) under C.c.

The Sun in the 12 Hours of Day

The sun in the 12 hours of the day according to the representations from the epoch of the Greco-Roman Egyptians.

Directory of the names of the foregoing twelve sun disks

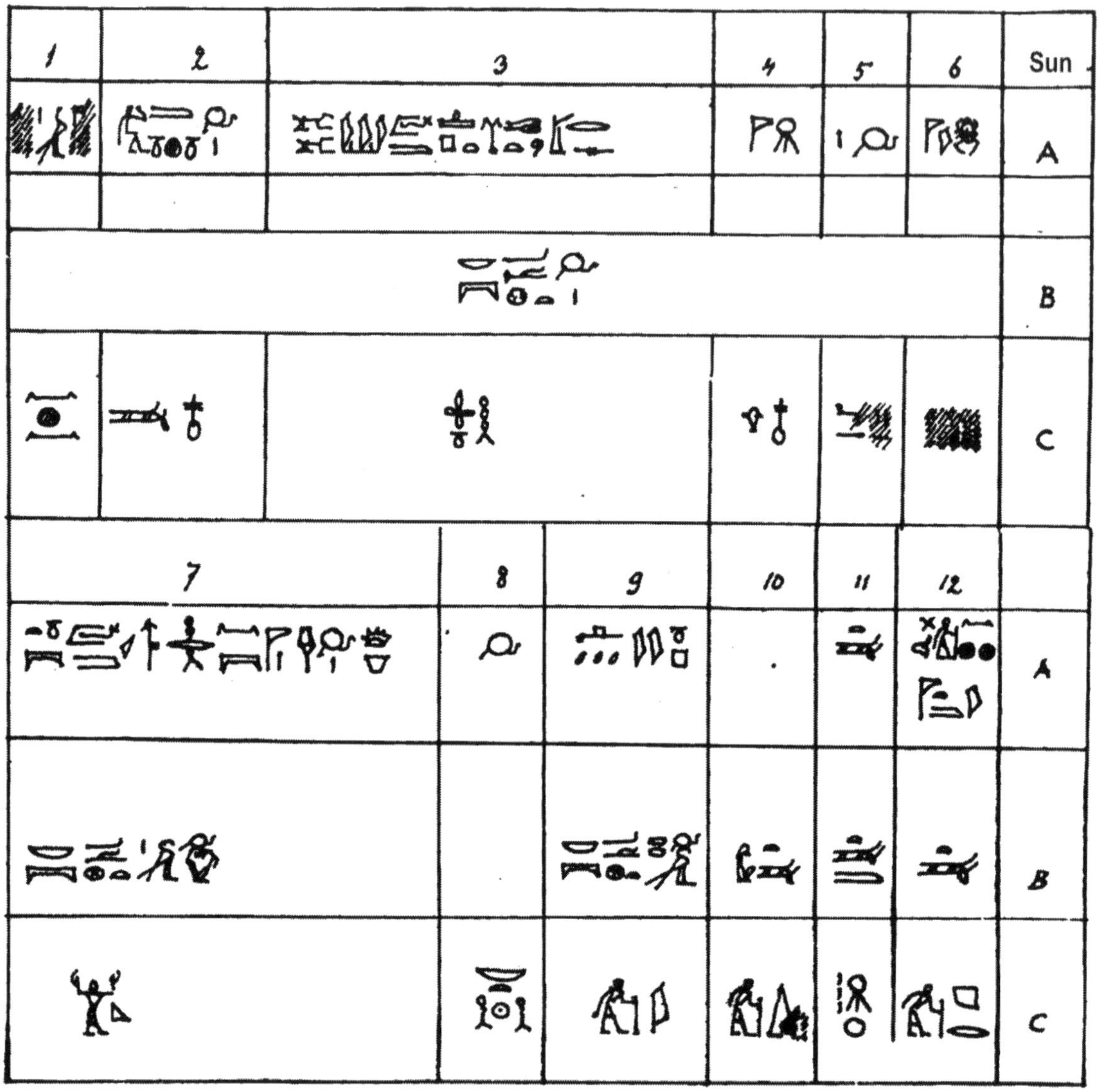

From the above list of the names of the suns in the 12 daylight hours, there is the fact, which is also confirmed by the other monuments, that the early sun as a child (nexen) in the 3rd hour (see C) as a young man (hun), and the evening sun as an old man (see C and A), and in the 12th hour of the day as an ancient man (nexx uer A.) It also emerges from this that the first nine hours of the day were designated as the sun  Ra, while the last three were designated as  atum. Like the confused succession of the sun boats, the draftsmen of the relevant representation in the *Description de l'Egypte* have in the same way confused the arrangement of the groups of figures, which are in the two outermost columns above and below the sun ships. The groups listed below, starting from the west, open the bottom row. At the same time, they are the only ones that have been added.

1. 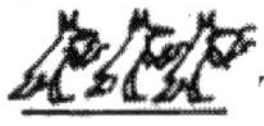 Text: 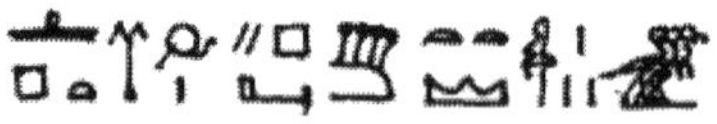 "The western spirits greet the sun as it goes down".

2.  Text: 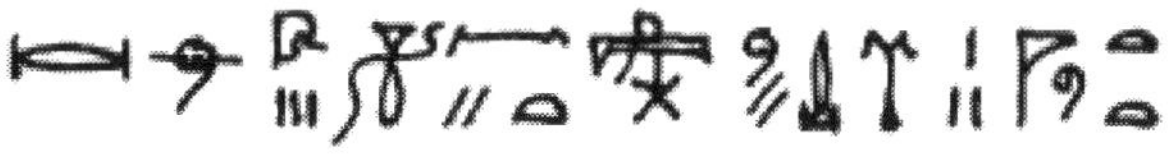 "God's gathering in the evening time which ... [what follows is incomprehensible]

3.  Text: 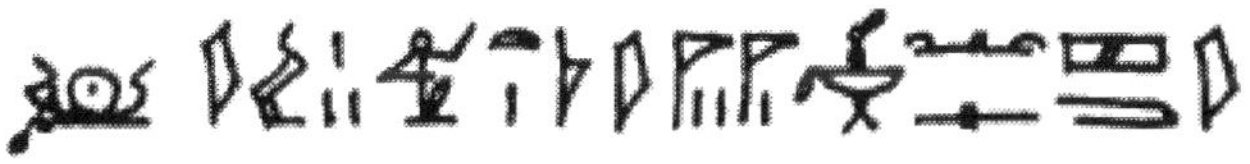 "The *asem-sek* gods (stars), the sailors of the sun ship".

4.  Text: 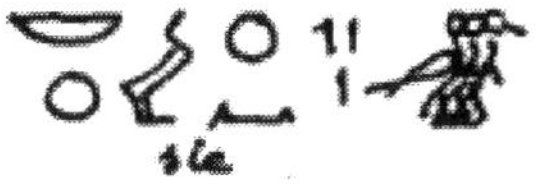

"The spirits on the first day of the month".

5.  Text: 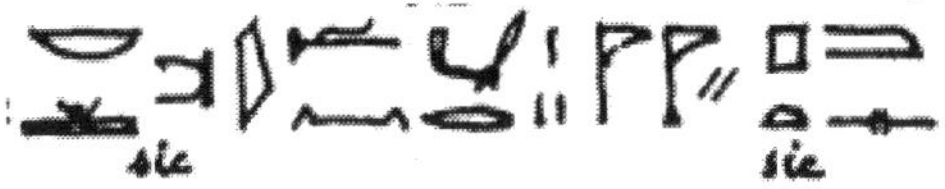

"the *Masepti* gods, the glorious ones, of the 7th day of the month."

———————————————————

With this, the astronomical ceiling paintings in the pronaos of Dendera have come to an end. I now turn to other astronomical images and inscriptions, which are also to be found in the aforementioned temple.

The Inscription next to the Zodiac of Dendera

As is well known, on the ceiling of the second room in the southern Osiris temple on the roof of the great Hathor sanctuary of Dendera, there is the world-famous circular image of the zodiac, which the French had expounded in order to set it up in the National Library in Paris. The inscriptions below belong to this zodiac, which can still be found today on the spot next to the former place of the zodiac:

Translation

"He, the glorious, addresses his words to the god *Osiris*, the Lord .... heaven every month, whose body is rejuvenated anew, whose will is higher than that of the god of heaven and that of the ruler of the earth, the *Orion* star that rises every day without disturbance on the body of the heavenly goddess: your divine form grows on the day of the new moon and you become a youth in the moon. It follows that the *Sauti* star as the leader (king) of the decan stars in your name of *Osiris-Orion*. Your sister, the shining *Siern* (i.e. *Isis-Sothis*) gives the direction of your run, echoing far away from you. Grant the honor of Sothis to your son, the king, for ever and ever!

The Position of the Sun on the First Day of the Normal Sirius Year

according to the ceiling picture
on the southern side of the zodiac room of Dendera (see page 50)

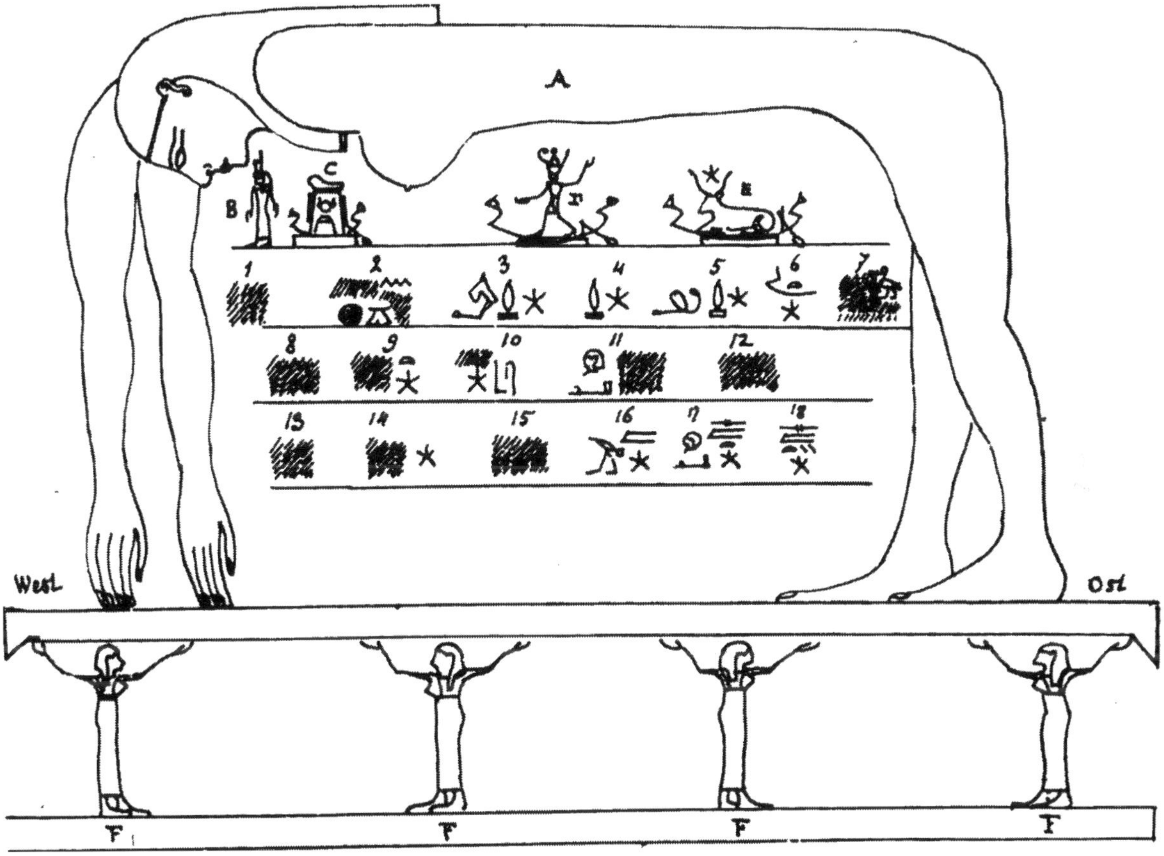

Explanation and translation of the inscriptions A. the *celestial figure* B. *Isis* i.e. the star *Sothis*-Sirius. C. the rising sun D. the constellation of Orion (Osiris-Sahu) E. the Isis cow as *Sothis*-Sirius star, therefore probably B. as Isis in the meaning of the year (proven by the monuments!). Summary 1-18 the first half of the *decan constellations*: namely:

1. [*Knum*, ΧΝΟΥΜΙϹ], 2. [*Xar-Knum*, ΧΑΡΧΝΟΥΜΙϹ], 3. *na-ta*, ḤΤΗΤ, 4. ta 5. phu-ta ΦΟΥΤΗΤ, 6. tum, ΤΩΜ, 7. [uśte, ΟΥΕϹΤΕ], 8. [bikoś, ΒΙΚΩΤ], 9. [Apos] ot, αφο=αφοϲο, 10. Seb[śos], ϹΟΥΧΩϹ, 11. tpā-[Xont], ΤΠΗΧΟΝΤΙ, 12. ......, 13. ......, 14. ......,15. ......, 16. Seśem, ϹΕϹΜΕ, 17. tpā-smat, (ΤΠΗϹΜΑΤ), 18. smati, ϹΜΑΤ.

F, F, F, F the four *heavenly* carriers as representatives of the four main directions of the *heavenly* regions.

The Moon

Position on the first day of the normal Sothis or Sirius year according to the ceiling picture next to the one described above

Explanation of the pictures below the celestial figure. A. Figure of the god Thoth, the Egyptian Hermes-Lunus, who stretches out his hands to B. the full moon, which seems to float on a pillar. Behind it, enclosed in a chapel, is the disk of the rising sun. Below these, the picture of a turtle. According to the decan's lists, the dean constellation called 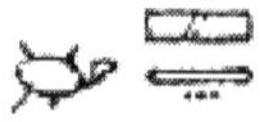 *seta*, i.e. "turtle", corresponding to its rising in the immediate vicinity of the Sothis - or Sirius star, takes its place between the end and the beginning of the normal Sirius year, as the compilation of the (more recent) decan lists (below) will clearly demonstrate.

Under F. the figure of *Osiris-Sahu* i.e. of the Orion star, next to it 9. (Isis cow lying in a boat) the very common idea of the Isis star of Sothis (Sirius). On the *fourteen steps of the staircase* (D) are the same fourteen deities that are shown and described in more detail above, page 32. They also appear in the image just discussed as the representatives of the *fortnight of the waxing moon*. On the bottom step of the staircase, the goddess *Anet* (or *Anti* i.e. the Hermonthian, from its cult in the city of *An*, *Anu* Upper Egypt, i.e. called Hermonthis) appears, who appears as the ruler of the *14th lunar day*, on the top right behind the image of the divine Thoth, the figure of the (Hermonthic solar) god *Mondθ* or *Monθu*. The border position indicates without any difficulty the beginning of a normal Sothis year when the full moon has entered. In the following list I have compiled the constellations of the *five planets* in the earlier and later periods of Egyptian history, as they have been suggested on individual monuments up to the present day. Here is the directory of the same:

A. Planet names with explanatory additions and and images in the sarcophagus room of *King Seti I* (XIX dynasty) in the valley of the king's tombs in Thebes.
B. Directory of planets with additions and illustrations in the astronomical ceiling picture in the temple of *King Ramses II* (the so-called Ramesseum) on the west side of Thebes (XIX. Dynasty).
C & D. Corresponding registers from king's tombs no. 5 and no. 9 of Bab-el-meluk from the times of the *XX dynasty*.
E. List of planets, next to the pictures, from the pronaos hall of the (Ptolemy) temple of *Edfu-Apollinopolis magna*.
F. Images of planets and names according to the images in the pronaos of the temple of Dendera (Roman period, see above p. 6 and following).
G. The same from the same pronaos (see above p. 4 and following)
H. The same from the round picture of the zodiac of Dendera.
I. The planetary images and their (hieratic-demotic) drawings on the lid of the wooden coffin of a *certain Heter*, son of a *Horsiese*, of Theban origin (see my *Recueil*[14] I pl.XVII and the accompanying text p. 30): Roman Epoch.
K. The demotically written planet names on the *Stobart planetary tables* published and explained by me, from the times of the Roman emperors. Location Thebes.

Other occasional mentions of planet names will be discussed at the bottom of the planet tables.

[14] Translator note: *Recueil de monuments égyptiens*

I Planet Table of the XIX and XX Dynasty

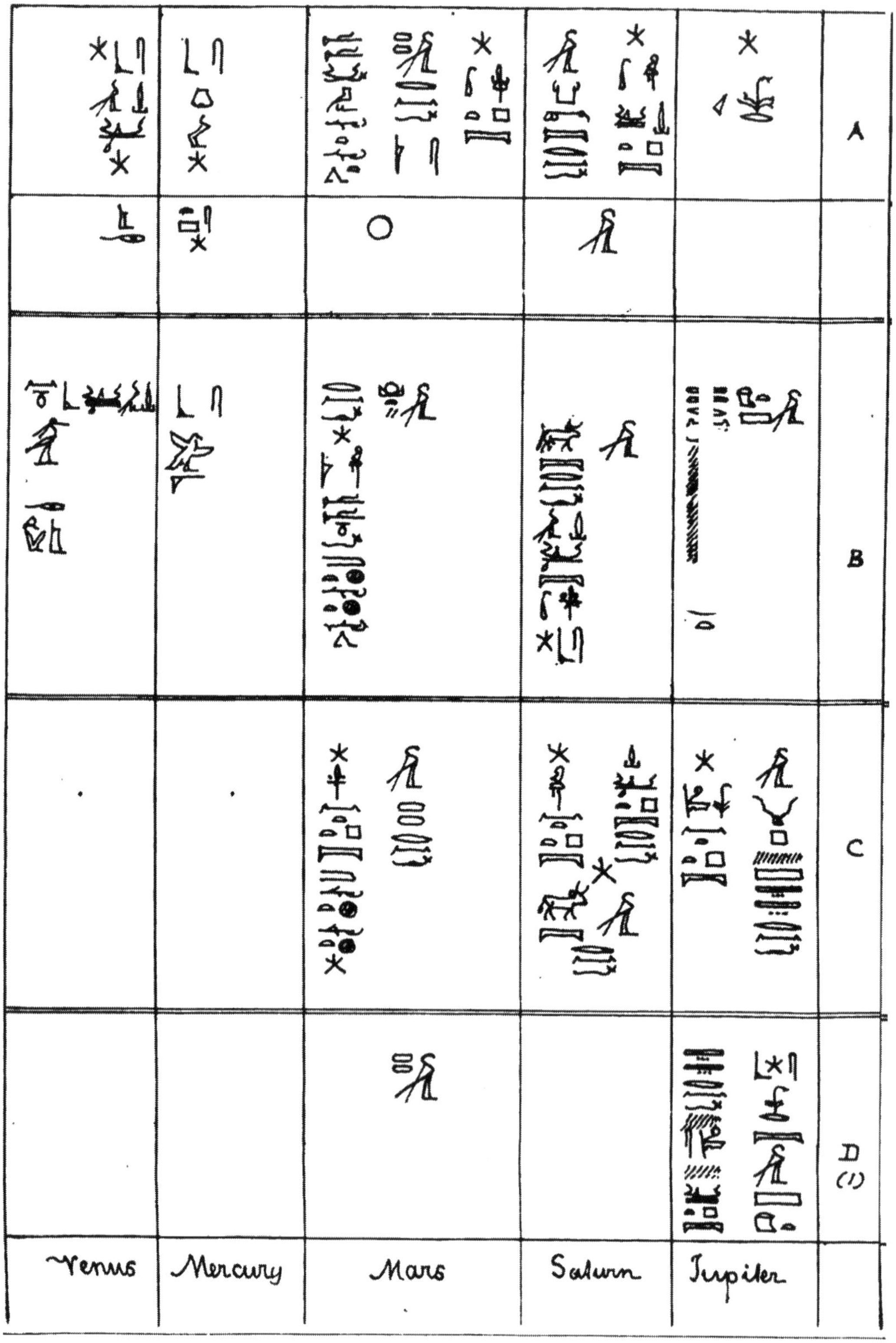

[Table continued from page 54]

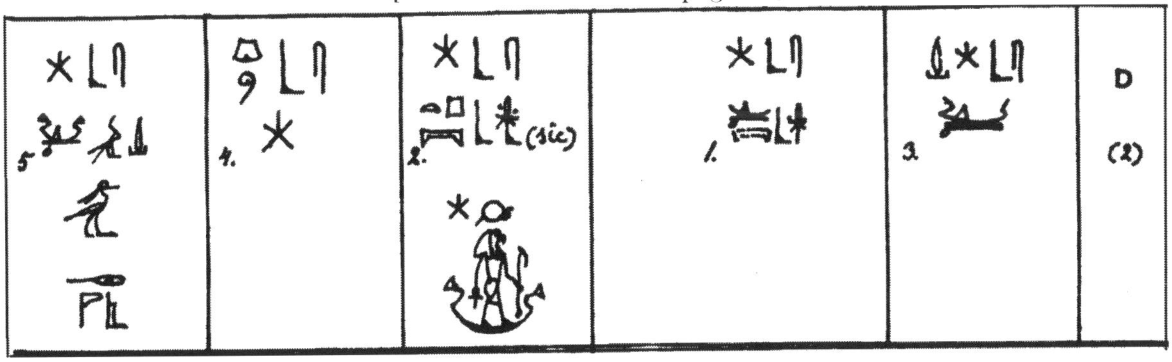

Planet Pictures

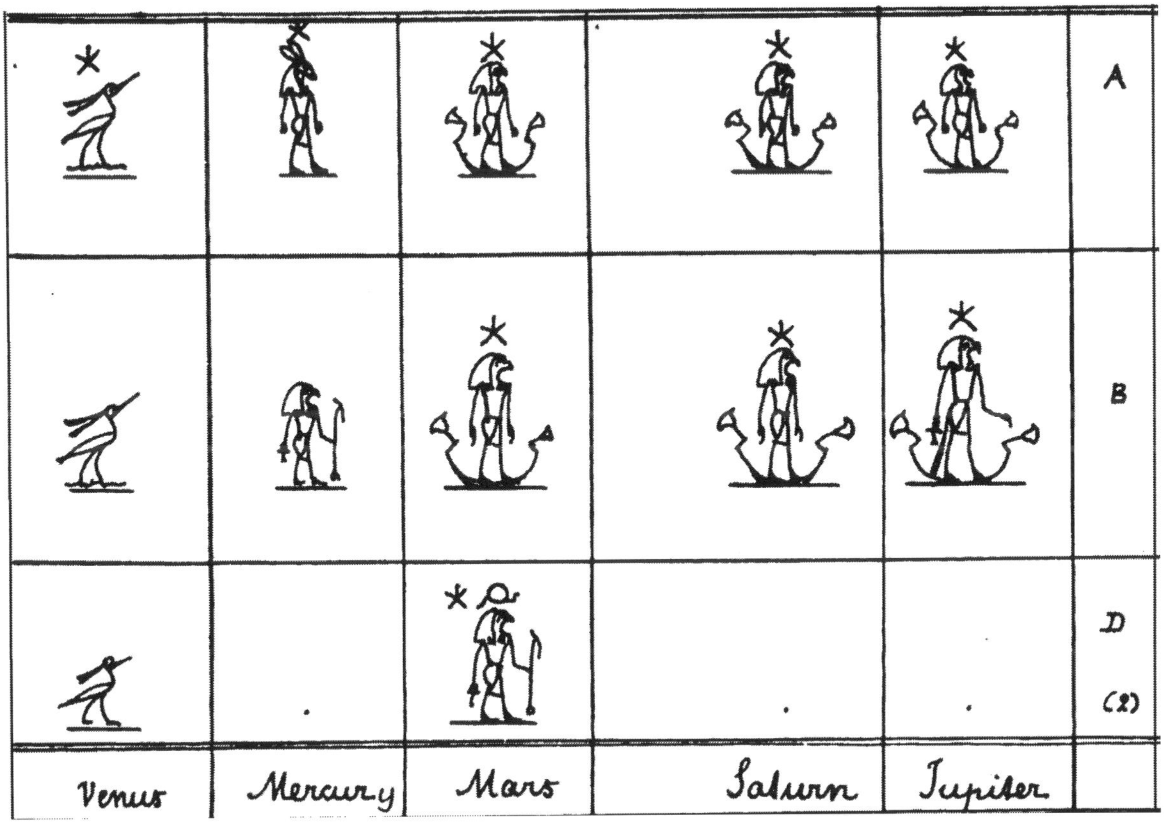

Note. The order of the planets in names and characters is the same in registers A, B, and C. In D (2), on the other hand, the arrangement, as indicated above by the appended numbers, is as follows: 3, 1, 2, 4, 5.

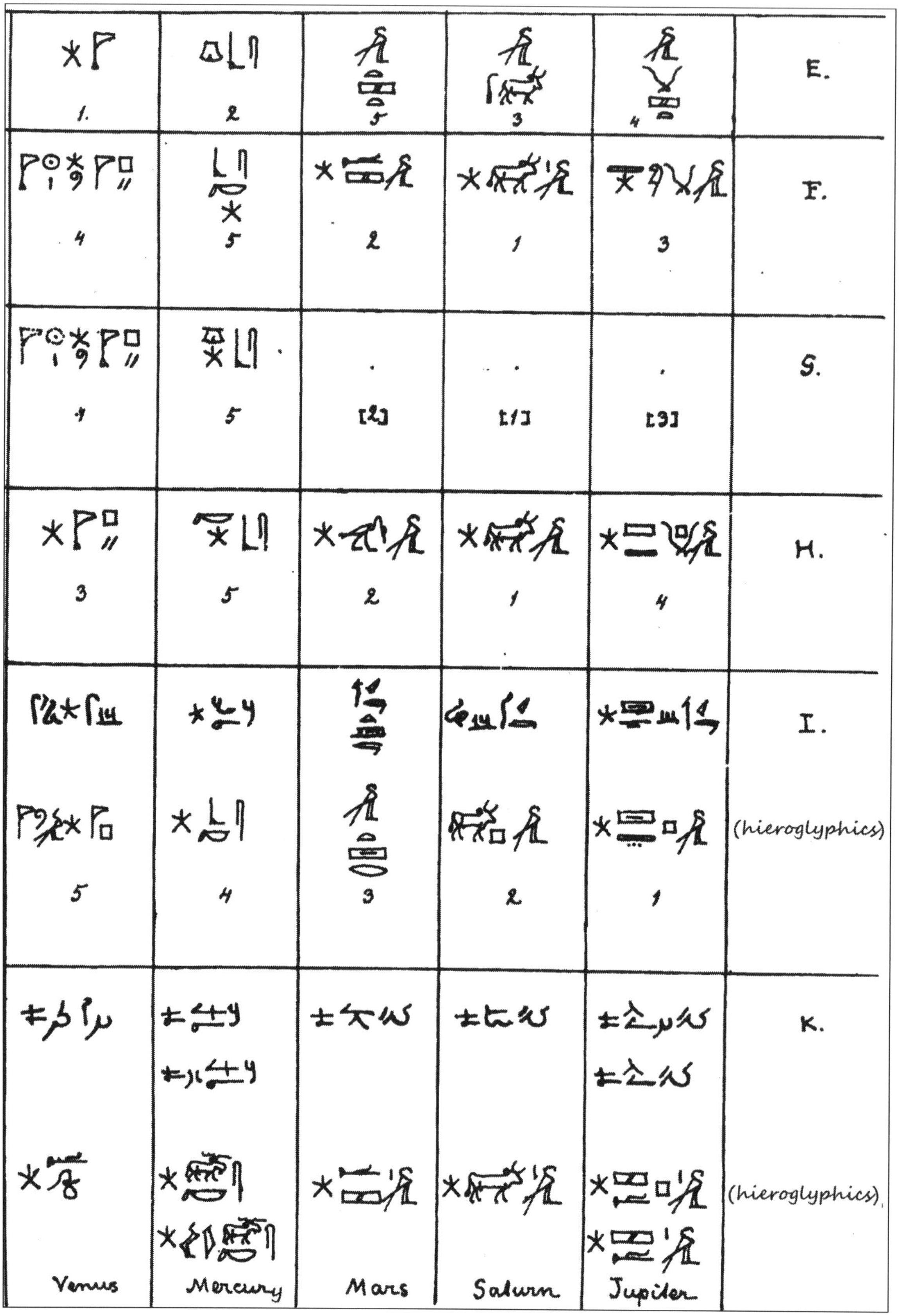
1.
2
5
3
4
E.
4
5
2
1
3
F.
4
5
[2]
[1]
[3]
S.
3
5
2
1
4
H.
I.
5
4
3
2
1
(hieroglyphics)
K.
(hieroglyphics)
Venus
Mercury
Mars
Saturn
Jupiter

II Planet Table of the Greco-Roman Era

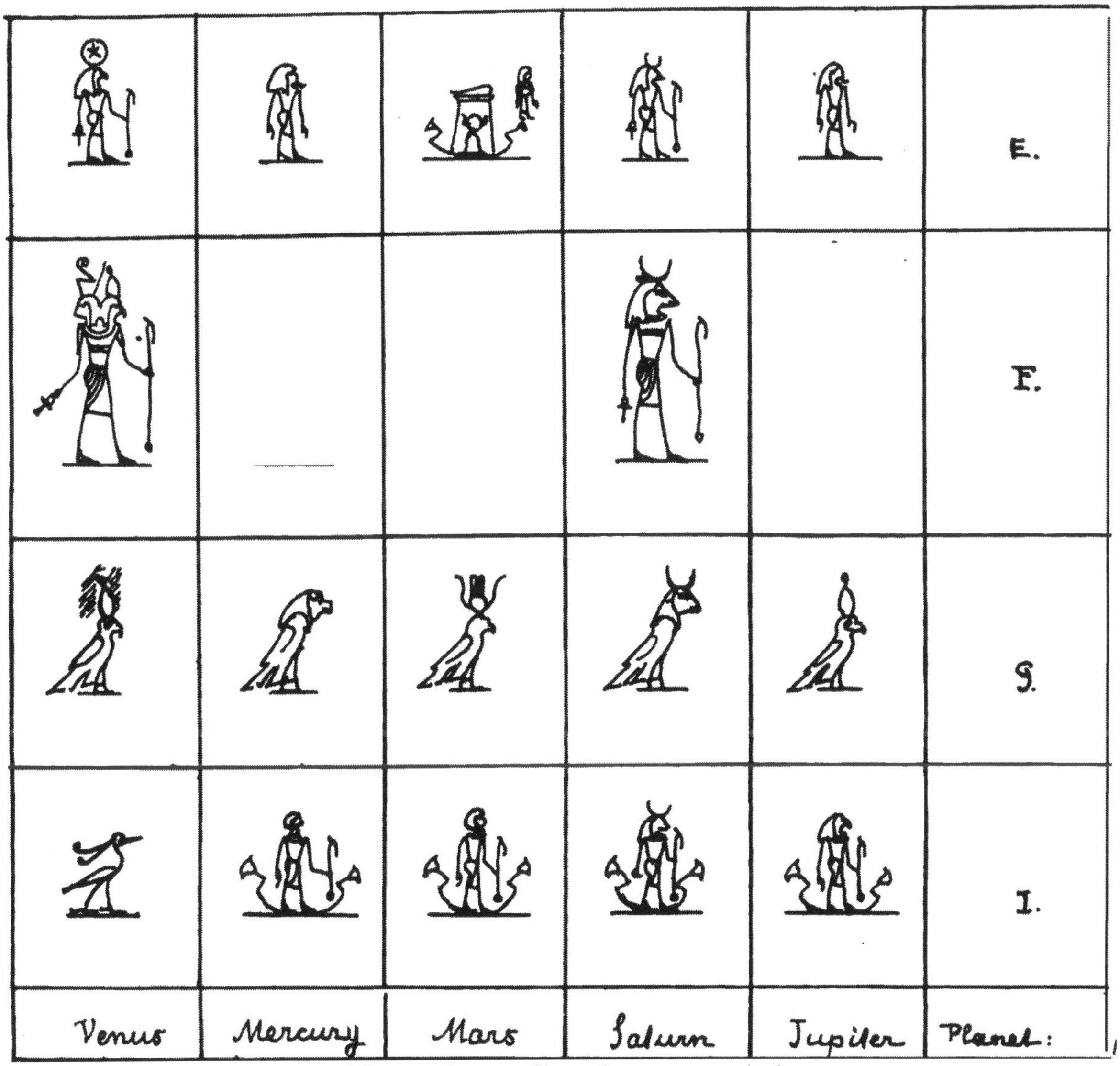

Planet pictures from the same period

In the following transcription of the above planetary registers from earlier (I) and later (II) times, I have made the comments relating to each individual planet follow the vertical order in order to make the overview of the various designations clearer and to facilitate their understanding.

1. Planet Jupiter. I. "The star of Ludens" (A) - *Hur-tas-ta* is called "he [the walking star of heaven in the] south" (B) - "*Hur-up-seta* is his name "The star of the south of the sky" (C) - "the star of the south of the sky is called *Hur-seta-ta* [the star... ..] walking through the sky (D, 1) - the walking star" - II. *Hur-up-set* (E) - *Hur-up-seta* (F) – ibid. (9) - *Hur-pe-seta* (I) - *Hur-pe-sed*, *Hur-sed* (K).
2. Planet Saturn. I. "The western star passing through the sky, it is called *Hur-ka-pet* (Horus bull of the sky)" (A) - *Hur-ka-pet* means "it, passing through the sky, the star is to the east"

(B) - "passing through the sky, it is called the western star of heaven. His name is *Hur-ka-pet*" (C) - "Star of the east, passing through [the sky] "(D, 2). II. *Hur-pe-ka* (θ?) (E) - *Hur-ka* (F, H) - *Hur-pe-ka* (I) - *Hur-ka* (K).

3. Planet Mars. I. "The eastern star of the sky, *Hur-xuti* (i.e. shining Horus) is his name. He goes through his path upwards" (A) - "*Hur-Xuti* is his name (the star) of the west. It goes through its orbit going backwards" (B) - "*Hur-xuti* is his name, the star in the east of the sky in reverse motion" (C) - *Hur-xuti* (D') - "the eastern star of the sky" (D2) – II. "*Hur-teset* (i.e. *Hor die* red, understood and represented as a woman) (E) - *Hur-dos* (Horus the red", F) - *Hur-dost* ("Horus the red", H) - *Hur-tost* (I) - *Hur-dos* (K).
4. Planet Mercury. I. *Sebqu* (A) - *Seb* ... (B) - *Sebqu* (D2) - II. *Sebgo* (E) - *Sebek*, *Sebko* (F.H) - *Sebgo* (9) - *Sebkau*, *Sebko* (K).
5. Planet Venus. I. "the walking star of Osiris" (A) -"a walking bird *Bennu* (a species of heron, Phonix) – Osiris" (B) – "the walking star bird" *Bennu*-Osiris" (D2) - II. "God of the morning" (*nutar dua*) (E) - "the god of the morning" (*pe-nutar-du*, F, G, H) - *pe-nutar-duau* (J) - *pe-nutar-du* (K) with the same meaning: "the god of the morning". The main differences between the later and earlier periods in the drawing of the planets concern the third (Mars) and fifth (Venus). In the former, Mars, , *Hur-xuti* was called "shining Horus", i.e. it had the same name as the sun, in this one it was called "red Horus" *Hur-dosr* or *Hur-dos*, a name that has been clearly preserved in the Greek Αςτηδ or Ερτωσ (see Lepsius, introduction[15] p. 90) and once (5, above E) is related to a female Horus form. The older drawing of the planet Venus as the star of Osiris or the Bennu bird of Osiris is replaced in the later period by "God of the morning". The representation of the same under F, from Dendera, shows him double-headed, on the right (left shoulder side) with a sparrowhawk's head and the crown on it, on the left with a human head and the north crown on it, no doubt an allusion to Venus as the morning and evening star. According to the older lists, the planets are consecrated to the following gods (see above, page 54, A, and page 55, D):
   1. the planet Jupiter (without specifying the water).
   2. the planet Saturn to the god *Hur*, Horus-Apollon,
   3. the planet Mars to the god *Ra*, Helios,
   4. the planet Mercury the god *Set*, Typhon,
   5. the planet Venus the god *Usiri*, Osiris.

According to Greek and Roman traditions, to which Lepsius (introduction p. 90) has already referred, the Egyptians should have considered: Jupiter as the star of Osiris, Saturn as the star of Nemesis, Mars as the star of Heracles, Mercury as the star of Apollo and Venus as that of Isis. The list from earlier times just given in no way agrees with this. Whether it is based on a different god scheme of the later monuments, at least one fact seems to support this. On the walls of the temple of Dendera, (e.g. in the Osiris sanctuary on the roof of the same) Osiris is referred to as king by the double name: 1. *Hru up-seta* 2. *Usiri-xont-amenti nutar a hir-ab Anet* ("Osiris, the one in the west, the great god in Tentyra"). The former name, in more recent spelling, corresponds to the same elements which form the

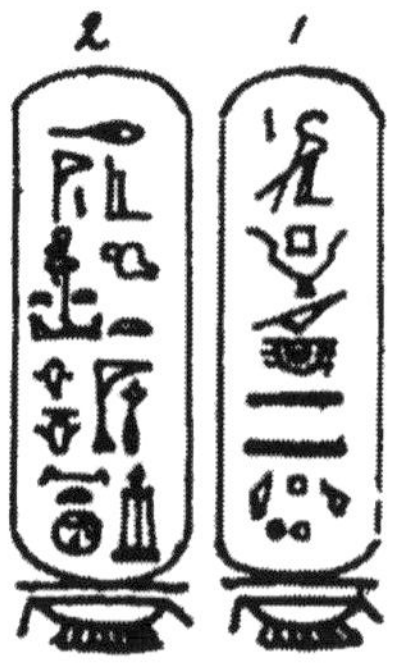

[15] Translator note: This probably refers to *Die Chronologie der Aegypter.*

planetary name *Hur-up-sheta* for the planet Jupiter. *Osiris was therefore actually conceived as Jupiter, in accordance with the Greek tradition.*
The above list of planets according to their Egyptian names and images was allowed to provide interesting contributions to historical astronomy. It is also expanded and enriched by inscriptions and texts which occasionally list planet names in connection with mythological ideas. I direct my attention above all to the following:

The Planet Venus

I. In the *Book of the Dead*, chap. 109, with the heading  *ro en rex biu abti*, "Chapter for knowing the souls of the Easterners" can be found, for example, in the Leiden Papyrus T, 16 (section LIII) the following vignette:

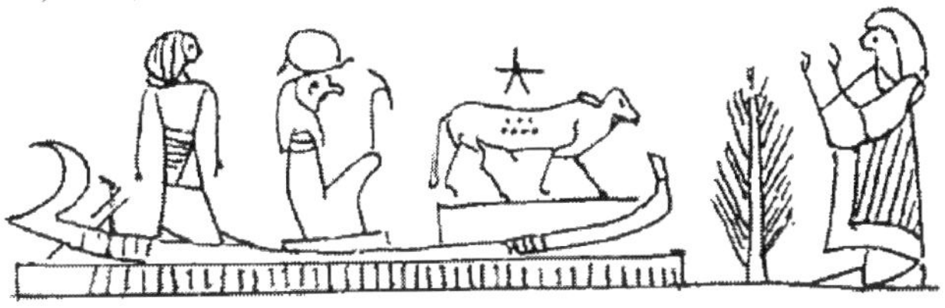

The related text puts the following words in the mouth of the deceased: "I know that eastern mountainous region of the heavens whose south is by the lake of *Xaro* and whose north is by the river of *Ro*, at the place where the day-god *Ra* sails in under gale-force winds. I am a welcome companion in the ship, and I row without resting in the barque of *Ra*. I know that tree of emerald green between which *Ra* appears when he walks over the cloud layer of the god *Su*. I know every *Thoth* from which *Ra* emerges. I know the realm of *Alo* whose wall is made of iron. His grain is 7 cubits high, its ears are 3 cubits and the stalks 4 cubits, according to the proportion of the spirits (the dead), each of which is 8 cubits in length. They sickle (the grain) near the eastern spirits (among the gods). I know the eastern spirits, namely the god *Hur-xuti*, the calf next to this god, and the god of the morning." The original text of the underlined:

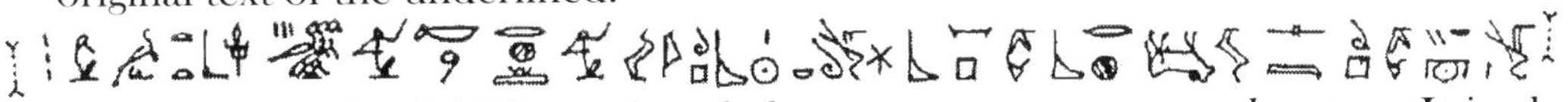

au-a-rex-ku-a biu abti Hur-xuti pu behsu xer nutar pen nutar duaut pu. It is clear that the spirits of the eastern region are considered to be:

1.  *Hur-Xuti*, that is, the planet *Mars*,
2.  *Behsu* "*the calf*", constellation in the east of the sky, and
3. 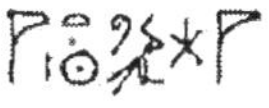 *Nutar duaut* the planet *Venus* or "*divine morning star*".

The east or morning region is naturally peculiar to the latter star. The relationship of *Mars* to the east is proven by the legends started above:  "the eastern star of the sky, which is called *Hur-xuti*" (see A) and 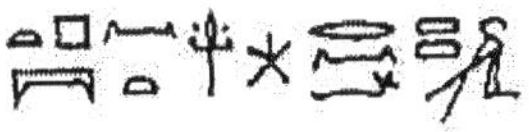 "*Hur-xuti* is its name, the eastern star of the Heaven" (C)

and  "the eastern one, Star of heaven (D[2]) next to his picture in a ship:

II. It seems to me the question of whether and when the ancient Egyptians already knew and established the identity of the morning and evening star cannot be answered beforehand. I seem to be able to prove by striking examples in my *Worterbuch* (Suppl., p.1358) that they have a special name for the evening star, namely, , female, 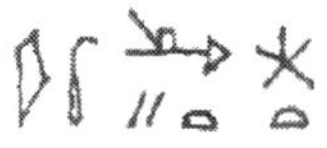 *sib uati* "the lonely star." Because of this, I quote the following text from Edfu (inner side of the western wall of the temple), in which the god Horus of Apollinopolis Magna is praised as the morning and evening star at the same time:

Hymn to the morning and evening star (see page 61).

(1) "The morning star, whose throne is in the firmament of the stars, who illuminates the earth with the light of his eyes.

(5) "The [star in] the west of the country *Punt,* which rises in the evening in the west of the sky and whose glow spreads over the earth

"The god *Horus* from *Apollinopolis magna,*

"The great God and Lord of Heaven, he is who the opponent falls to the ground on the Baltic ropes every day."

"He is the is the evening star which one thanks for looking"

(2.6) *Hymn of praise to this God*

Hail to you!"

"You morning star!"

"You evening star!"

(3.7) "Hail to you, Horus!"

"You standing upright."

"Who rises."

"Hail to you, *Horus* of *Apollinopolis magna*, you great God and Lord of Heaven!"

"Whose rising out of the primordial water is life and revelation,"

"You who fall to the ground, the dragon *Apophis* in the east of the firmament in heaven on earth, in the waters, on the mountains"

"that they (sic) no longer lift up their head for eternity: all enemies of the fall to the ground;"

"that they (sic) never roused each other again: leave your glorious face straight."

"Son of the sun, the king, in heaven, on earth, in the water and on the mountains!"

"Be to the son of the sun, to the king Ptolemy, to the everlasting living, to the friend of Ptah!"

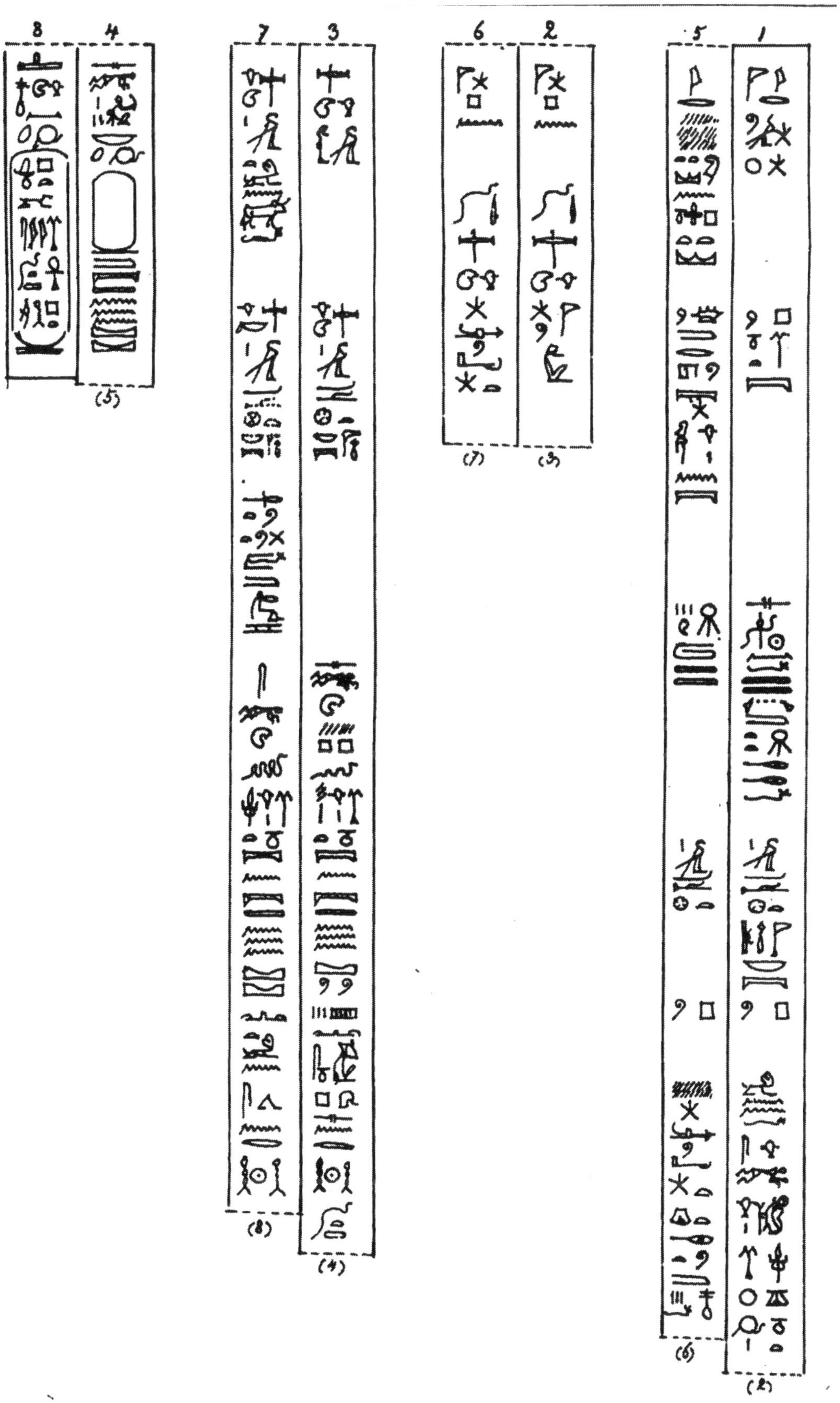
8
4
(5)
7
3
(8)
(4)
6
2
(7)
(3)
5
1
(6)
(2)

In the face of such a clear and distinct example, any doubt as to the meaning I discovered of the word 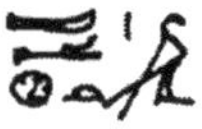 sib mat or mati in the sense of evening prayers must vanish. It also emerges from the same inscription that the god Hur-bahudti or Horus of Apollinopolis magna, the Egyptian Apollon, in one certain aspect enumerated the planet Venus as morning and evening star. This was illustrated figuratively by the idea in Edfu, shared by Naville (*Mythe d'Horus* pl. 19), to which the above text refers.

On the south side (, not , as Naville erroneously says) the god mentioned is symbolically denoted as  planet Venus shining in the west, and correspondingly on the north side (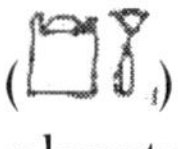) as  planet Venus shining in the east. The close relationship of the same god to this planet is also sought elsewhere, namely

III. by the following call, which Thoth addresses to this deity in the text of the birth and struggles of Horus from Apollinopolis magna to Edfu: 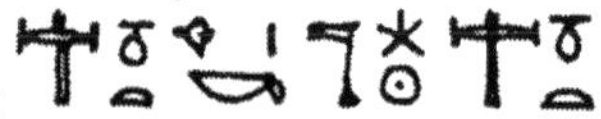

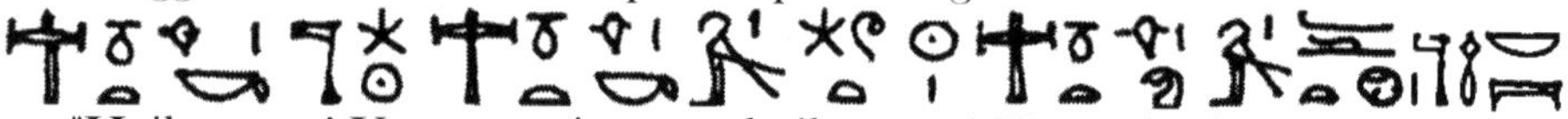

"Hail to you! You morning star, hail to you! Horus in the morning, hail to you! Horus of Apollinopolis magna"

IV. This view of Horus as the planet Venus explains the meaning of the following mythological genealogy, based on the inscriptions of Dendera and Edfu:

 Ra-Hur-Xuti i.e. the sun god

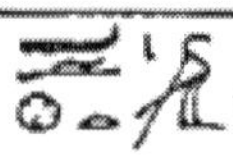 Hur-behudti i.e. Planet Venus 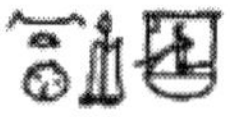 Hathur neb ant i.e. Sirius

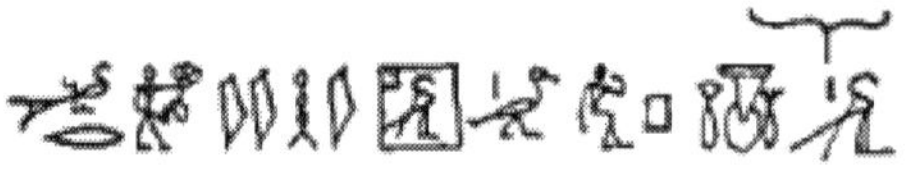 Hur-sumta-pe-Xrud si Hathur ahi uer

i.e. Hursamta the child of the son of Hathor, the noble (or: older) Ahi i.e. the young New Year's sun (see above p. 8 line1 ff.)

In astronomical terms, the content of this mythology is only this: the (Sothic) New Year's Day occurs when, at dawn, *the sun, the planet Venus and Sirius become visible in the sky at the same time*. Incidentally, according to the written traditions,  was understood as the sun in Upper Egypt. Compare: 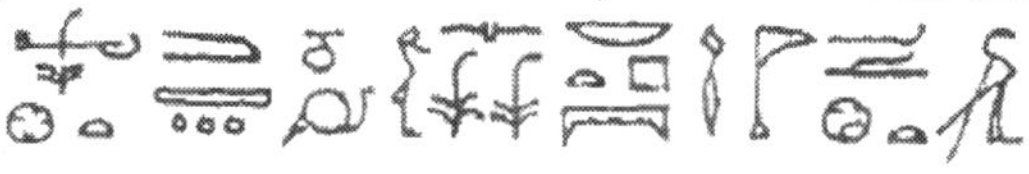 "Horus of Apollinopolis magna, the great God and Lord of Heaven, *the figure of the sun in the land of Upper Egypt*" (ta-gemat) [NMH,[16] 13] and likewise:

[16] Translator note : Naville. *Mythe d'Horus.*

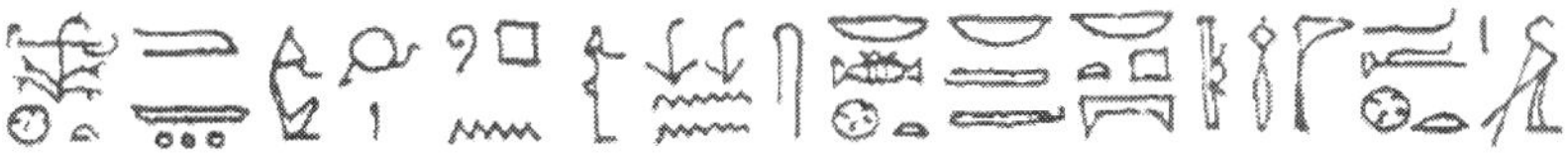 "*Horus* of *Apollinopolis magna*, the great god, the lord of heaven, the lord of the earth and the lord of *Masenet* (name of his sanctuary in Edfu) that is the *figure of the sun in the land of Upper Egypt*" [DTI[17] I, II, t].

The Planet Jupiter

I have already established its relation to *Osiris* above. The same planet name of god (*Hur-up-seta*) recurs in following places: 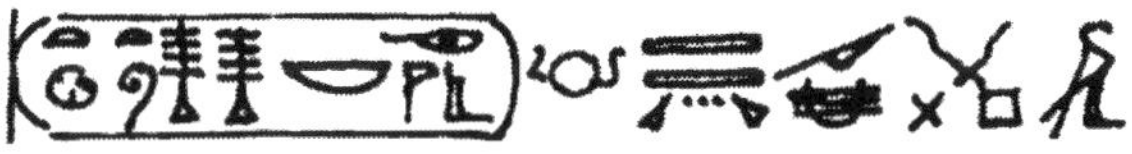 "*Hur-up-seta*, a sun, the *Busiris Osiris*." [MD.[18] IV, 36, 40]. 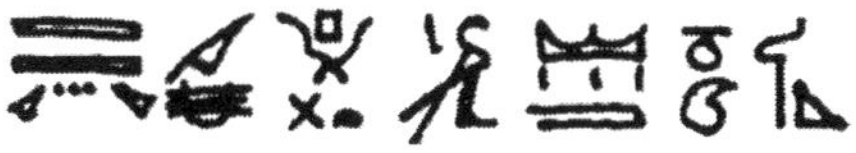 you are king of the people as *Jupiter*" (*Hur-up-seta*, ibid. 64).  *Hur-up-Seta* i.e. appears in a series of emblems. *King Jupiter* (ibid. 21. IX). On a stone Osiris statuette of the *v. Huber*'s collection the god is called: 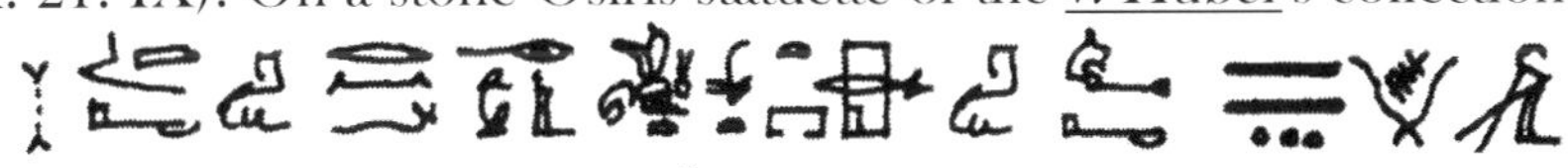  "*Jupiter*" the Prince in the palace (of Heliopolis), the king of Upper and Lower Egypt *Osiris*, whose real name is *Uonn-nofir* (Onnophris), his father is *Seb* (Kronos), his mother is *Nut* (Rhea), his city is named *Thebes*." In the sub-Egyptian city of 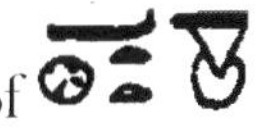 *Sambehudet* i.e. the Jupiter city of  *Hur-upt-seta* i.e. "King Jupiter" (see my *Dictionnaire Geographique,* page 1380).

The Planets as Suns

That the planets, and with them a few other stars distinguished by their shimmering glow, were regarded as *suns*, is proven by the monuments. If *Horus* of *Behud* or Apollinopolis is called: 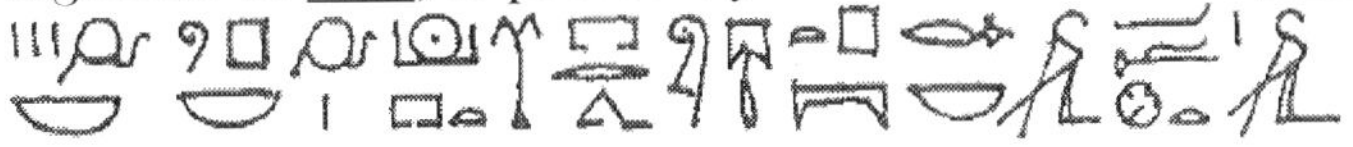 "Horus of Behudet, the great god and lord of heaven, the brightly plumed, who rises from the dwelling of light, that is namely "*the sun of the lord of the suns*" (*ra pu neb ra neb*) [BWI.[19] LVI, 1], as the final sentence shows that the term *ra* not only referred to the *daytime* sun, but that it also referred to certain *shining stars*. So begins an inscription I copied in Philae with the words:

---

[17] Translator note: Dümichen. *Altägyptische Tempelinschriften*
[18] Translator note: Mariette. *Dendera.* Vol. IV
[19] Translator note: Brugsch. *Wort.*I

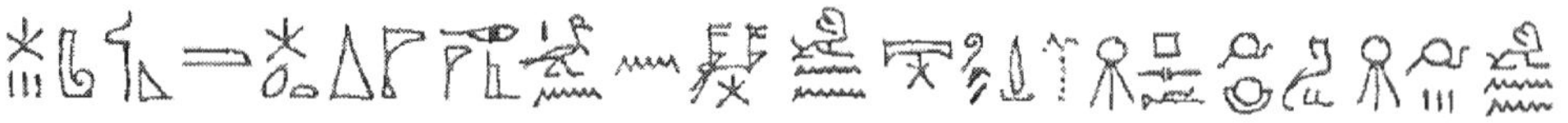 "They are sun, which shine daily and which shine in the twilight, they are (these) the *Sahu* (Orion) stars of the soul of *Osiris* and the *Sothis* (Sirius) stars as the king of the decan's stars." In fact the goddess is frequently called *Hathor-Isis* of Tentyra (i.e. *Sothis*) in the inscription: 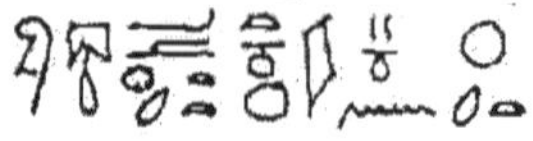 "the second [feminine] sun next to the solar disk, the *Apollonian* [goddess], the one with colorful feathers" [DBD.[20] 56/1]. Similarly, from the planet tables (see the same under A and D[2]), it emerges that the planet  is clearly defined as  or  i.e. *Ra* "sun" was cited. Likewise, the planet name of *Osiris-Jupiter* listed above on p. 63 line 2 is connected with  *Ra* "sun." Similar examples are not uncommon in the written traditions.

The Constellations of the Sky

Besides the planets, the monuments play a big role. I shall follow them in representations and inscribed descriptions as they appear listed next to and one after the other from the monuments of earlier and more recent times. As the most important and masterful, at the head of the ages are *Sirius* and *Orion*. See the next page.

[20] Translator note: Dümichen. *Bauurkunde der Tempelanlagen von Dendera*

The Constellations of Orion [*sahu*] and Sirius [*Sopedet*]

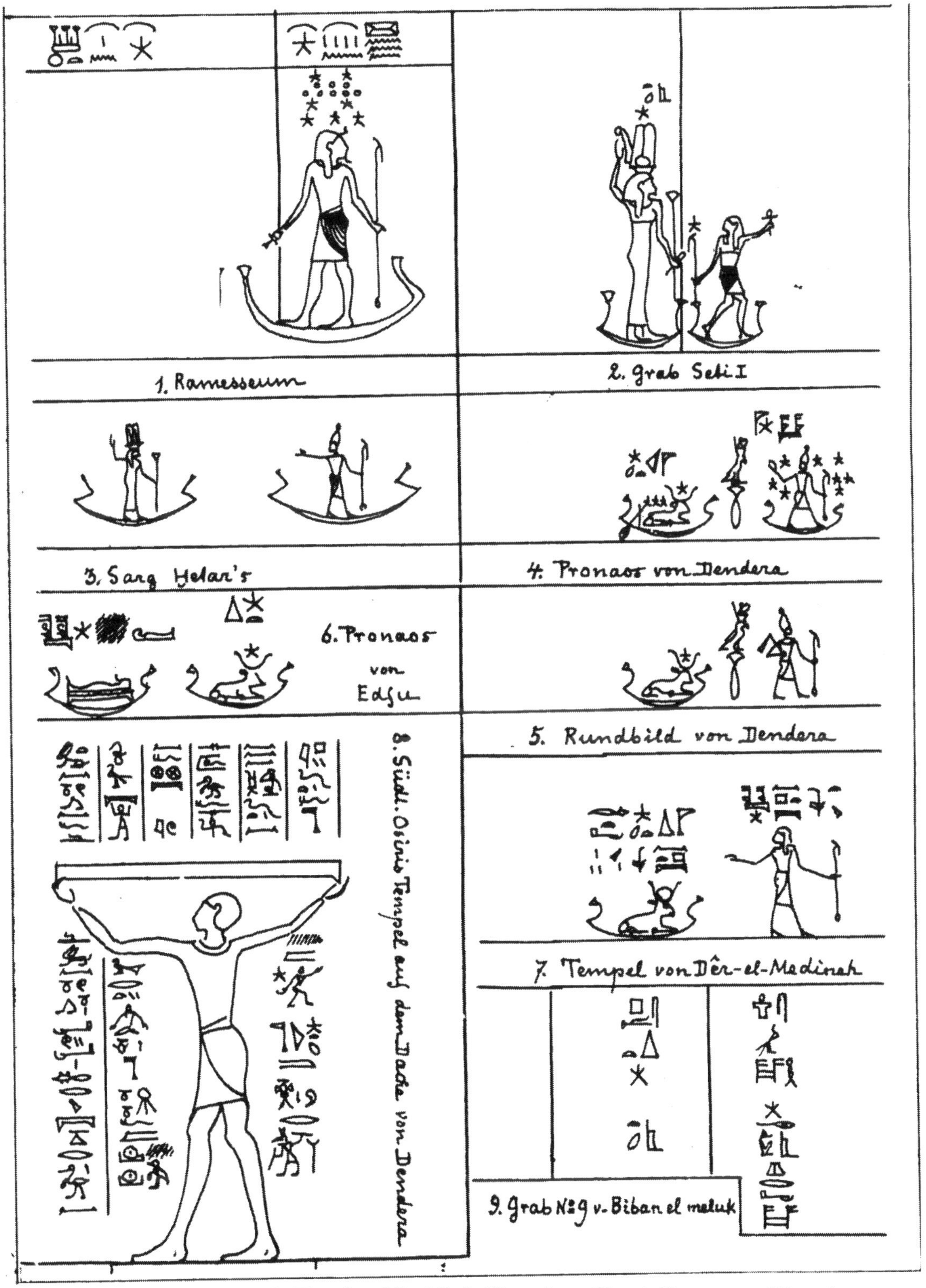

1. Ramesseum, 2. Tomb of Seti I, 3. Hetar's coffin, 4. Pronaos of Dendera, 5. Round picture of Dendera, 6. Pronaos of Edfu, 7. Temple of Der-El-Medinah, 8. Southern Osiris Temple on the roof of Dendera, 9. Tomb No. 9 of Biban el Meluk.

The Constellation of Orion and Sirius

In the older, and in the majority of the more recent representations, the deities of *Osiris-Orion* and *Isis-Sothis* appear in barges, surrounded by stars, which indicate that they did not refer to a single star, but to a group of stars, to a sign of the zodiac (constellation).

Orion is shown as a running man in the guise of a king, with his face looking back and averted. His name already appears in texts of the *sixth dynasty*, such as in the inscriptions of the pyramid of King *Hun-em-saf* (or *Sokar-em-saf*, as others read the same name). It is called  *sh, sah.* A fuller spelling shows  *sah"* (see above No. 9) with the addition:  (*sah* star) "of Osiris under the hand of *Sah.*" Abbreviated spellings of the same name are 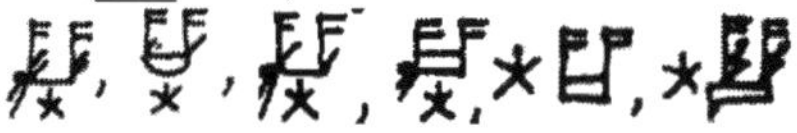 and others. The word is undoubtedly based on the stem  *sah*, whose meanings have been partially preserved in the later Coptic ⲥⲁϩⲉ, with suffixes, ⲥⲁϩⲱ *avertere*, *amovere*.

The *Sah* star cluster referred to as *one who turns away*, according to the pictorial representation. Other inscriptions recognize in the same name the word 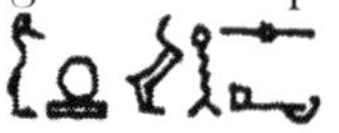 *Sahu*,  or 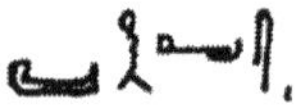, 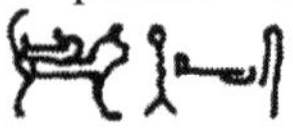 *sah* "the mummy." The same representation is repeated in Esne, where Orion himself and Sothis are shown in a flying manner:

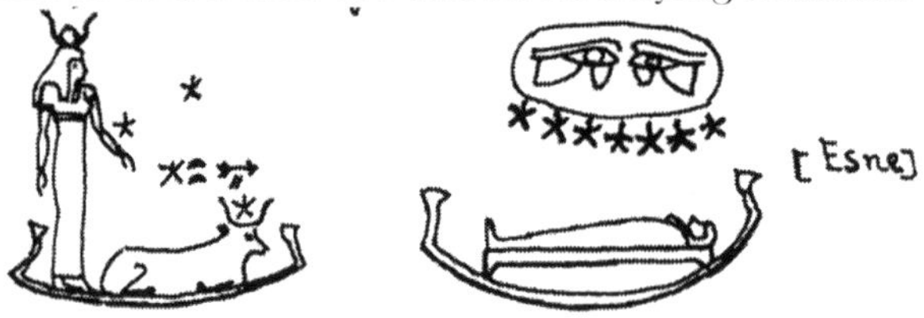

while at the same time Orion appears a little to the left above the Isis-Sothis barge in the form shown here.  I derive from this that Orion was conceived in different guises depending on his position in heaven at the main times of the year, at least from the end of the Ptolemy rule onward (refer to Edfu). According to Egyptian sources, in the 21st chapter of his treatise on Isis and Osiris, Plutarch remarks: "The priests say not only the body of Osiris, but also the bodies of the other gods, not eternal and not immortal, lay with them after death and were venerated, *the souls* (τὰς ψυχάς) *shone in the sky as stars*, (ἄστρα); so the *soul of Isis* is called by the Hellenes *Hundstern* (κύνα), by the Egyptians *Sothis* (Σῶθιν), *those of Horus* (sic) *Orion* (Ὠρίωνα), those of *Typhon* the *bear* (ἄρκτον)." In the 22nd chapter the same writer remarks: "They consider the ship called by the Hellenes *Argo* (Ἀργώ) to be an image of the vehicle of Osiris, which is placed under the stars out of reverence, not because it is distant from Orion and the dog stars; *that one*, they think, dedicated to *Horus*, *this one* to *Isis*."

These statements of the excellent Greek writer are confirmed by the inscriptions of the monuments, but with a necessary correction: not the *soul of Horus*, but the *soul of Osiris* is called *Orion* (*Sah*) by the Egyptians. Compare the text published above under No.8 for the following content. The words of the heavenly bearer are put in his mouth: "I have raised my two hands that carry heaven with the soul of the god *Uerti-ab* (i.e. whose heart has stood still," common nickname of the god Osiris). He rises in the place of light as a *Sah*-star (Orion, ). The divine Sothis () is protective behind him in order to chase away the enemy." The 6-line text above it reads: "He has risen and he supports heaven to let his (Osiris) soul enter. He enlightened (made bright, made shining, made shine, see Plutarch above) the physical form of the god *Api-tet-f* (, new epithet of Osiris). Now one should compare the inscription given above on p. 12,[21] which confirms the expression  *bau, biu* "*souls*" for the stars. On the west side of the frieze in the first room of the Osiris temple on the roof of the great Hathor sanctuary of Dendera it says of Osiris:

 "His soul has become divine under the stars, always rising (or "living," since the verb *anx*, , has this double meaning) as a *Sah*-star (Orion) on the body of the sky goddess *Nut*." Also on p. 7 with the inscription (no. 31) it reads: "Orion, the magnificent soul of the god Osiris."

According to the text from *De-el-Medineh* (see p. 65 No. 7), Orion is also called 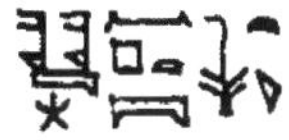 *Sah en pet sut* "the Sah star of the *southern sky*." As well as Sirius, or rather the whole constellation of the same: 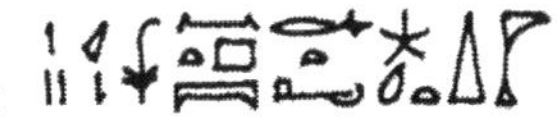 "the divine Sothis, the great one of the *southern* sky." The position of the constellations mentioned in the southern sky gave them, and especially Orion, the meaning of the *South*, in contrast to the constellation of the  *mas-xet* "leg" (see below) in the *northern* heaven, representing the northern region, the *North*, and belonging to the god *Set*-Typhon, the great bear or chariot. Compare: 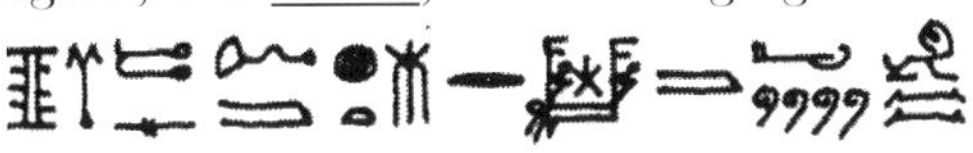 "There are 400 cubits from Orion (i.e. from the south) to the Great Bear (i.e. the *north*) on the parapet." [see my *Dict. geog.*[22], page 1396].

[21] Translator's note: located in column 1.

[22] Translator's note: *Dictionnaire géographique de l'ancienne Egypte.*

In a building inscription that has already been discussed by me elsewhere (in duplicate A and B), the following words are spoken by a Ptolemy at the foundation ceremony of the temple of Edfu:

i.e. A. "the stretching of the measuring cord
B. "the stretching of the measuring cord at the temple between 2 pegs.

A. "Speech. I have grasped the stake and the handle of the hammer. I take the measuring rope in communion with the goddess <u>*Safex-abui.*</u> I consider the forward movement of the stars. My eye is on the Great Bear. I count the time, check the clock and fix the corners of your house of worship."

B. "Speaking. I have grabbed the stake, I grasp the handle of the hammer, I take the measuring rope in communion with the goddess <u>*Safex-abui.*</u> I direct my face to the course of the heavenly bodies, I let my eye enter the constellation of the great bear. There is the counter of the time next to his watch. I note the cornerstones of your house of worship." I will provide evidence below that as early as the epoch Ramses II Orion-Sah was ascribed to the southern sky 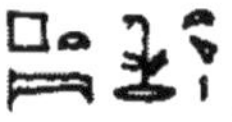. Otherwise noteworthy texts in which Orion is mentioned:

1. In the courtyard of the Temple of Edfu:

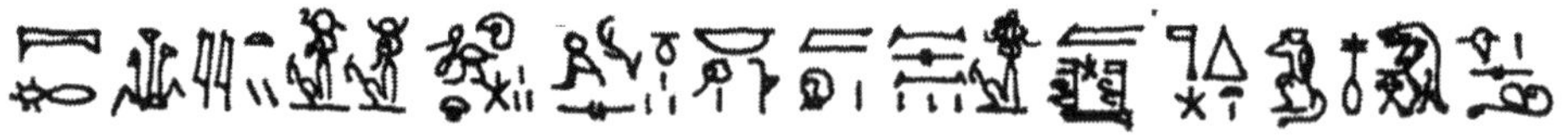 ....

"the sky (or "the roof" of the temple, the inner ceiling of the same) carries the two lightbringers <u>*sun*</u> and <u>*moon*</u>, the <u>*decans*</u> are behind them, <u>*the lord of the beginning of the year is first for them*</u>. Osiris as Orion, the divine Sothis, the good (or beautiful) <u>*baboon*</u>,[23] the <u>*hippopotamus*</u> -----. "

2. A lengthy text begins in the same courtyard:

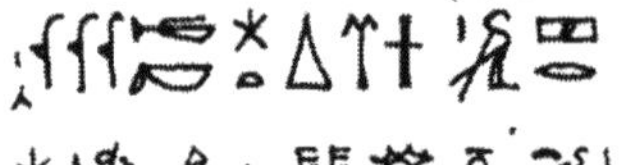 (Ptolemy Alexander) 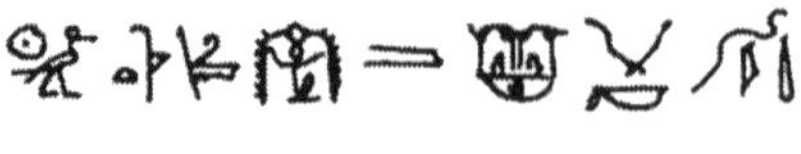

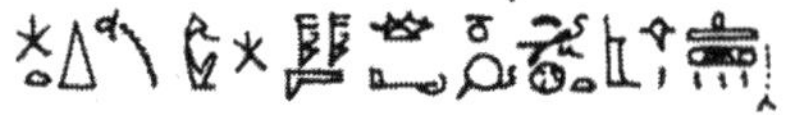

---

[23] <u>Translator note</u>: The German word *Affe,* used by Brugsch, is typically translated as *monkey* or ape, but we translate it as baboon. See <u>Appendix A</u> for the explanation.

3. On the long 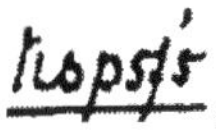, a gate shows the inscription:

Speech. The Osirian prophet of the god *moon*, the lord of Thebes, *nes-pesef*, "the triumphator," passes through the sky as *Sah*-star (with averted face). He is associated with the decan stars and one" (see Rec.[24] I, 70, 2).

4. On one of the posts of the coffin of *Hetar* (with the astronomical representations) the inscription: 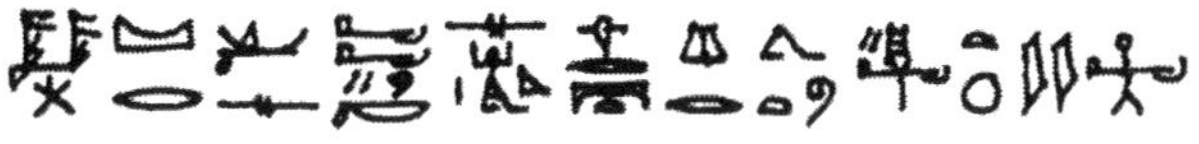 "The post stands erect and supports the heavens. You let your hands reach out to him (heaven) with the *Sah*-Orion," and similar other texts which are discussed below.

Isis-Sothis. Designation of the same in the inscriptions of *Hun-em-saf*'s pyramid (6th dynasty)  *sopdet*, fuller spelling of the same in the 20th dynasty (see page 65, No. 9)  meaning "triangle," usually *sopdet*, called "the triangle of god" (*sopedet nutar*) and written , 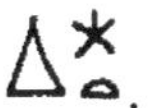. In the later (Greco-Roman) epoch the triangle was represented in the following form: 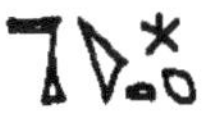 and sometimes 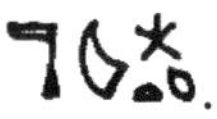. In the same epoch the older name *Sopedet* was often replaced by the more recent designation: 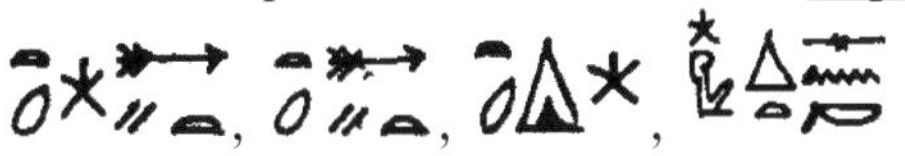 (see my *Worterbuch*, Sup. 1152) *satit* "the shooting one, protection" or "the swelling (the Nile)." From this, in all probability, the Greek name ⲤⲰⲐⲒⲤ emerged (Compare to Coptic ⲤⲀⲦ, ⲤⲈⲦ, ⲤⲒⲦ jacere, projicere, ⲤⲞⲦⲈ sagitta). More rarely, in the script, its designation by,  as in the following inscription from Esne: 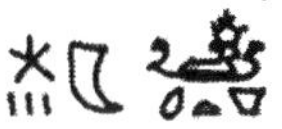 *satit hint xabsu* "Sothis queen of the decan stars."

---

[24] Translator note: *Recueil de monuments égyptiens.*

The following give the most detailed information about the meaning of this constellation.
1. The side inscription on the ceiling of the *astronomical hall in the Ramesseum in Thebes*:

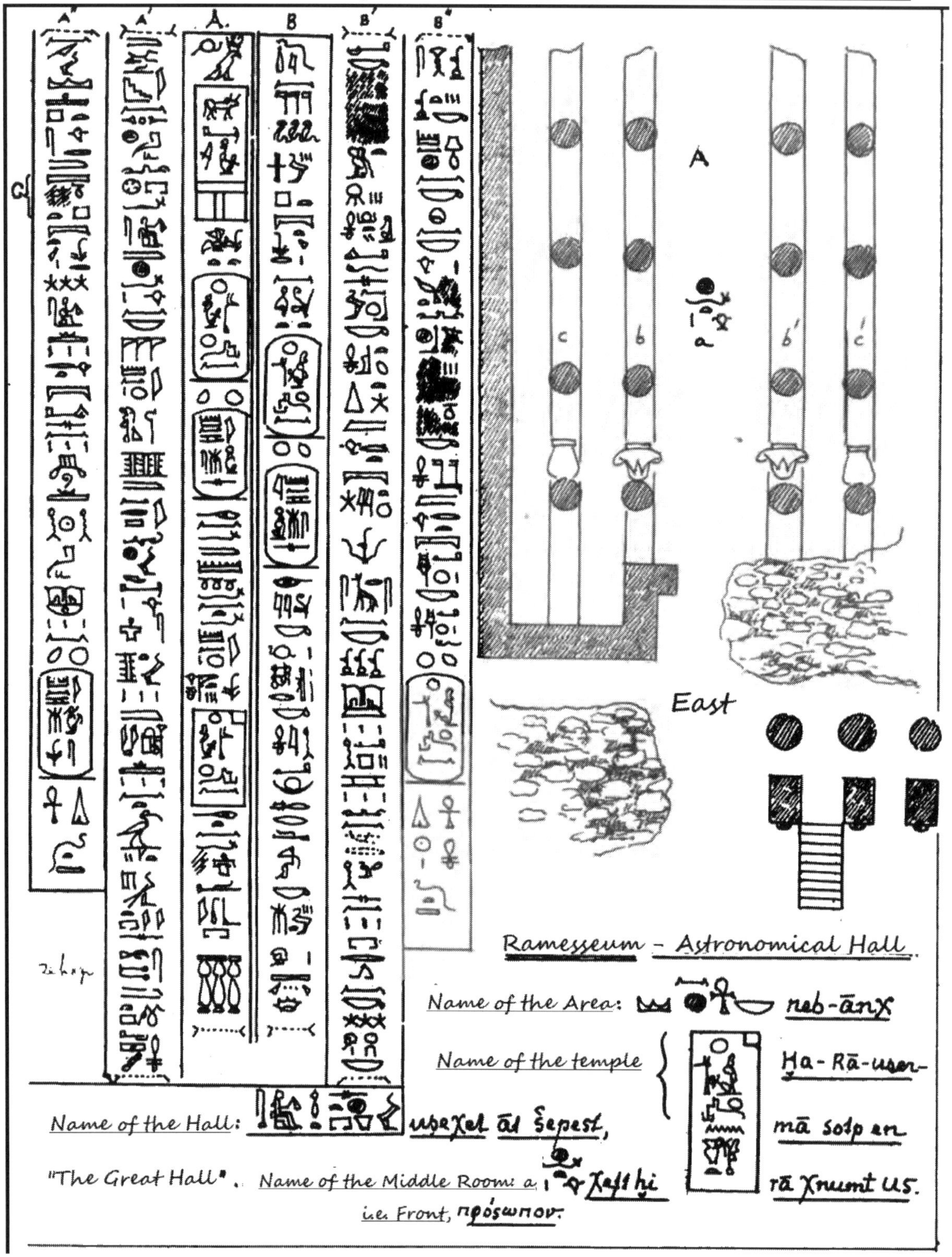

Architrave - Inscription to the north at b on the tarpaulin:

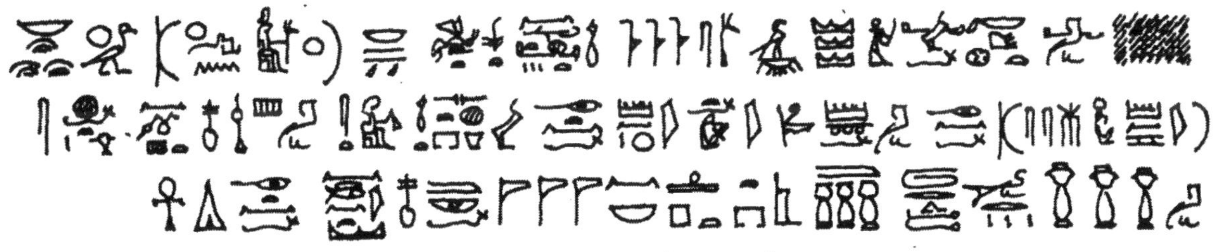

[Complete the missing beginning with: 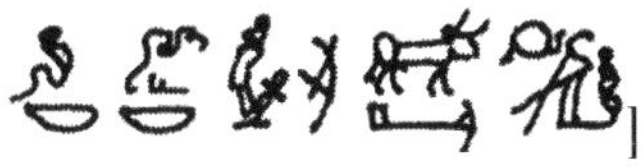]: The transcription of this dedication in its entirety reads thus: "*Ra-Horus*, the strong bull, the friend of justice,- t*he lord of the diadems*, the protector of Egypt and chastiser of the peoples, *Horus the conquerer*, rich in years, great in victories, *the king of Upper and Lower Egypt*, the sovereign *Ra-user-mat, sotep-en-ra the son of Ra and lord of the crowns*, *Miamon Ramses* (II), he has made donations as a sign of remembrance for his father *Amon-Ra*: the large magnificent hall made of light and good hard stone, whose central front consists of mighty pillars with flower capitates, which are bordered by columns with bud captiates, a resting place for the lord of the gods at his beautiful feast of the valley (*hib-en-ant*, in Coptic preserved as month name ⲡⲁⲱⲛⲉ, ⲡⲁⲱⲛⲓ, ⲡⲁⲩ̈ⲛⲓ[25] of the Greeks in the Alexandrian period.). He donated (it), the life giver."

Translation of the text [see A-B page 70]
in the astronomical hall of the Ramesseum in Thebes

A. "*Ra-Horus*, the strong bull, the friend of righteousness, the king of Upper and Lower Egypt *Ra-user-mat sotep-en-ra*, the son of *Ra Miamun Ramses*, he made his donation as a token of remembrance of himself for his father *Amon-ra* the king of the gods, in the *Ramesseum*: the hall of pillared columns to its splendid adytum with a central front for the lord of the gods *A-mon-ra*, the king of Thebes, the walls of stone with carvings, the walls arched according to the ordinances of the god *Thoth*, their entrance gate , made of hard stone, like (the arch) of an acacia tree from the land of *Setu*, and that which rests on it (i.e. the *roof* of the portico) with a replica (read C ) of the southern sky and the glorious stars of the heavenly heights. They give a long time in thirty year periods to the son of *Ra Miamon Ramses*, the giver of life today and forever.

B. "Thus speak the gods and goddesses in the *southern heavens* to the king of Upper and Lower Egypt *Ra-user-mat sotep-en-ra*, the son of *Ra*, *Miamon Ramses* your being is that of the sun and your becoming like that of the moon, your sweetness is that of those who are born on earth. You appear (among) men like the shining god. He lets you rise radiant like Isis, the Sothis star, the heavenly one in the eleventh hour of the night of the beginning of the year (i.e. New Year's Day). She promises you the end of thirty years of jubilation and inviolable floods of the Nile. You go to the stars at the beginning of every ten-day week (i.e. decade). The *New moon* is widening for you, not letting it decrease [your months.] You [appear like *Sah-Orion* in the sky. Your lifetime is the same as his lifetime, son of *Ra*, *Ra-user-mat sotep-en-Ra*, you giver of life, today and forever. "

---

[25] Translator note: ⲡⲁⲱⲛⲉ, ⲡⲁⲱⲛⲓ, ⲡⲁⲩ̈ⲛⲓ

Of particular importance in this text is the clause  "he lets you rise radiant like *Isis-Sothis* to the elevation of heaven *in the eleventh hour of the night of New Years Day*." I used to - and my successors with me - know the important group 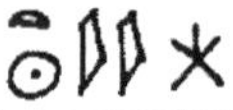 as a special variant of the well-known word for morning 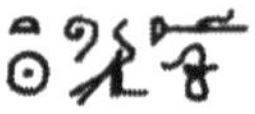 *duat*, 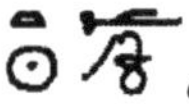 *duaut*,  *duat*, 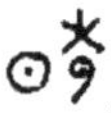,  *duau* [see my *Lex. Sup.* p. 1357) and translated accordingly 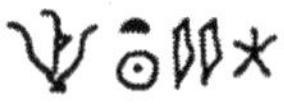 "On the morning of the new year's day" which is why passages in the inscriptions such as 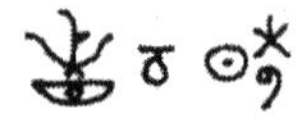, 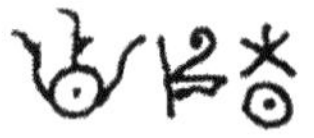,  *duaut en apet(en) (renpit)*, "the morning of the beginning of the year" frequently seemed to fit, only the different spelling of the word 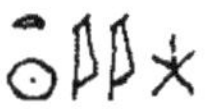 is immediately reminiscent of the designation , 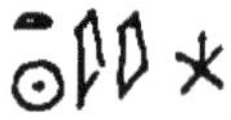, varr. 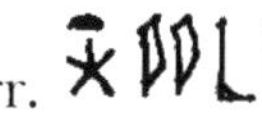, 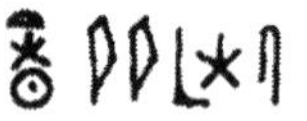 i.e. *sebit* which leads to the eleventh hour of the night on the monuments before the Greco-Roman epoch (see the hour tables below) Accordingly, the rising of Sirius took place in the eleventh hour of the night, in accordance with the Greek tradition:

ἡ τοῦ κυνὸς ἐπιτολὴ κατὰ ἑνδεκάτην ὥραν φαίνεται καὶ ταύτην ἀρχὴν ἔτους τίθενται καὶ τῆς Ἴσιδος ἱερὸν εἶναι τὸν κύνα λέγουσι, καὶ τὴν ἐπιτολὴν αὐτοῦ

"The rise of the *dog star* (Sirius) takes place around the eleventh hour and they regard it as the beginning of the year and they think that the dog star of *Isis is sacred* as is its rise" (Theon, in Scholl. ad Arati Phaenomena,[26] see my "*Materiaux*"[27] p. 100). In the same "*Materiaux*" on p. 103 I have provided evidence that, according to the information given, the star rising in individual royal tombs of *Biban-el-meluk preceded* the day, so that the first hour began with sunset and the twelfth ended just before sunrise. Hence the legend:

 *Month of Thoth, beginning of night*, ... *first hour, beginning of the year*. "The night of New Year's Day began in the first hour and preceded it, but the *actual beginning of the year* did not occur until the Sirius star rose at the *eleventh hour of the night*. The instructive example of this is provided by the hour tables in the tomb of King *Ramses VI*. Under the heading "*Night of the month Paophi*" 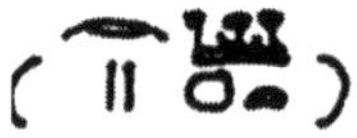 you can find: 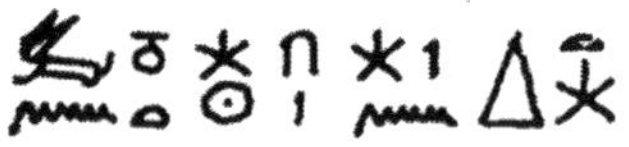 "*Hour 11, the star of Sothis*" (i.e. rises). Since the 1st Paophi falls 30 days later than the 1st Thoth or the beginning of the normal Sothis year, it necessarily follows due to the nature of the changing year, that the mentioned Sothis rise in the 11th hour of the night of the 1st Paophi 30x4 or 120 years later than the rise of the Sothis period from the year 1322 BC, i.e. *in the year 1202 before the beginning of our Christian era.*

[26] Translator note: Possibly Aratus, Theon, & Cicero, M. T. (1793). *Phainomena kai Diosemeia: 1.*
[27] Translator note: Brugsch. *Matériaux pour servir à la reconstruction du calendrier des anciens égyptiens*

If the year of the reign of King Ramses VI had been added to the timetable, in which the rise of Sothis on 1 Paophi had occurred, we would have been able to determine the beginning of the reign of the respective king with the most exact chronological certainty.

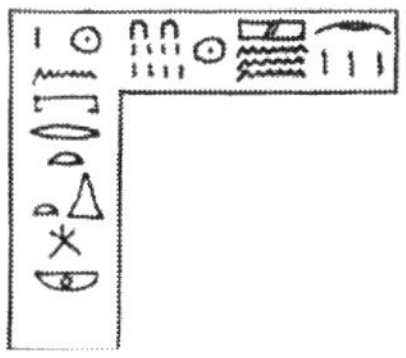

Another, much more favorable case concerns the time of the reign of King *Thutmose III* of the Eighteenth Dynasty. On the now famous calendar stone by *Elephantine*, which is now in the Louvre and belongs without any doubt to the government of the Pharaoh mentioned, there is a rising of *Sothis* indicated in the legends "*Month of Epiphi, T.28. the day of the celebration of the rising of Sothis* " The relevant year of the reign of the kings lacks this information. According to the inscription in *Amon-em-hib's* tomb in Thebes (see Zeitschrift, 1873, p. 7), *Thutmes* III ruled:

 "from the *year 1 to the year 54, Phamenoth last* (i.e. the 30th day). The day and month of his taking office are not listed, but can be supplemented by the relevant information in a fixed list on a wall in a sanctuary built by the king mentioned in the middle of the great temple of *Amon* at Karnak. One reads the following:  i.e. *"Month of Pachon, day 4, festival of the crownings as king of the king Ra-men-Xeper, of the everlasting."* This gives the most desirable confirmation of the same partially illegible date in the so-called statistical table of Karnak, from the same reign. The same, based on my own transcription, read as follows: *Year 23, month of Pachon, day 4, the day of the crowning as king.*" *Thutmes III* had therefore entered his government on the 4th of Pachon and ruled for 53 years, 11 months and 1 day until the day of his death.

In the same static table the dates preceding the year 23 Pachon 4 are related to the year 22 of the king's reign, e.g. the date immediately preceding year 22, month Pharmuti ......," in other words, there is evidence that *the kings had the annual accounts of their reign counted from the day of their accession to the throne*.

Other convincing datings from the same epoch make this proof incontrovertible, as one will conclude for himself from the individual data. In the hieratic papyrus No. 3226 of the Louvre, belonging to the 18th dynasty according to the character of the writing, dating calculations are presented with steles, whereby the 4 pachon is presupposed as the borderline of an old and a new year, as one can convince oneself from the following excerpts on the basis of my own copy of the long papyrus described on both sides.

I. Year 28, Pharmuti 10. a. Year 29 Pachon 6. – – Pachon 17.	Year 29 Pharmuti 17. b. Year 30 Pachon 8. – – Epiphi 5.	Year 31 Payni 27. c. Year 32 Epiphi 17.
Year 32 Pharmuti 24. d. Year 33 Pachon 5. – – Epiphi 4.	II. Year 28 Pharmuti 10. a. Year 29 Pachon 6. – – Pachon 17.	Year 29 Pharmuti 12. b. Year 30 Pachon 8. – – Epiphi 5.
Year 30 Phamenoth 25. c. Year 31 Pachon 21. – – Payni 13.	Year 31 Payni 28. d. Year 32 Epiphi 17. Year 32 Mesori 19.	Year 32 Pharmuti 24. e. Year 33 Pachon 5. – – Epiphi 4.
Year 33 Pharmuti 20. f. Year 34 Pachon 19. – – Payni 25.	III Year 29 Pachon 12 (sic) a. Year 30 Payni 30. – – Payni 19.	Year 31 Pharmuti 1. b. Year 32 Pachon 5. – – Pachon 21.
Year 31 Phamenoth 7. c Year 32 Mesori 19. – – Thoth 1.	IV. Year 28 Pharmuti 14. a. Year 29 Pachon 6. – – Pachon 10	Year 29 Pachon 12. (sic) b. Year 30 Payni 5. – – Payni 19.
Year 31 Pharmuti 2. c Year 32 Pachon 5. – – Pachon (?).		

The examples given will suffice to serve as proofs of how, even in the pharaonic administrations, the annual accounts were based on the date of the accession of the king of the period. Since under King Thotmos III a rising of Sothis is recorded on 28 Epiphi, the same (as the 328th day of the year) (366-328)x4 = 38x4 = 152 Sothis years (of 365 days) before the epoch year 1322 BC, i.e. 1474 BC. With a precise astronomical calculation of *the new moons* which occurred immediately before and after this date, the reign of King Thutmos III could be determined with absolute certainty compared to a calendar from his reign. We know that the king's ascent to the throne took place on the 4th Pachon. The celebration of the same was however transferred to the next new moon. Such is the case in the following text of the statistical table by Karnak [the copy I took myself]:

Renpit XXIII. Tep semu haru 21. haru en hib en paut er meti Xat suten tep duaut (sebit).

"In the year 23, month of Pachon, day 21, day of the *celebration of the new moon* festival, according to the correction of the coronation festival at the beginning of the morning (or: the eleventh hour of the night, see above)." The astronomical calculation of this new moon according to (Jul.[28]) year, month, day and hour would determine the date of the year 23, month of Pachon, day 21 to the greatest possible accuracy and thus offer the opportunity for the day and year of Sothis on the rise under Tuthmos III to be verified irrefutably.

[28] Translator note: Jul. = Julian?

Information about the Rise of Sothis Sirius in Mythological Conception in the Temple of Dendera

The *Sothis star* is understood as the *goddess Isis*, her *birth* represents the expression: *rise*, as her *mother*, the sky goddess *Nu* or *Nut* appears, the temple of Tentyra is her *place of birth*, more precisely the little *Isis Temple* located to the west of the great Hathor Temple. An inscription on the outer east wall of it (from Roman times) notes the following about it:

i.e. "*the life! the female Horus, the youthful one, the daughter of a* regent (*haq*), *Isis, the great one, the mother of the god, is born in Tentyra in the night of the child in his cradle on the western side of the temple of Hat-seses*" (in other words: of the great temple of Hathor).

The rising of the Sirius or Sothis star on the 20th of July (Julian) in the vicinity of the sun, or the heliacal rising of the same, is expressed mythologically in the same way. The *sun* is represented by the god of the day *Ra* or by the *boat* in which he ascends heavenward. At the same time, the presence of the morning star can be read from individual places. After these preliminary remarks, I have selected the most important texts, which are almost exclusively represented in the Temple of Dendera, in the following:

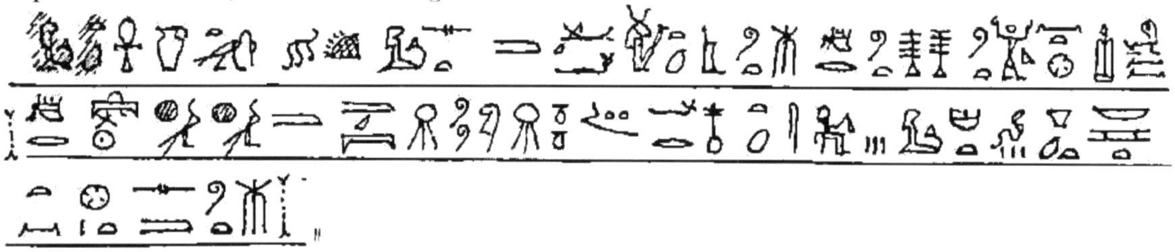

"It is the city of *Anet* in constant elevation when the (cow-headed) *goddess Isis is born* in it (the little Isis sanctuary) in the form of a dark red woman (called) *Xnum-anxet* (this is also the name of the 12th hour of the day in the older lists, just before sunset), the mistress of love, the queen of goddesses and women, the bride. You can already see the shimmering *rise of the light beam in the sky in the twilight, when she is born* in this city. "

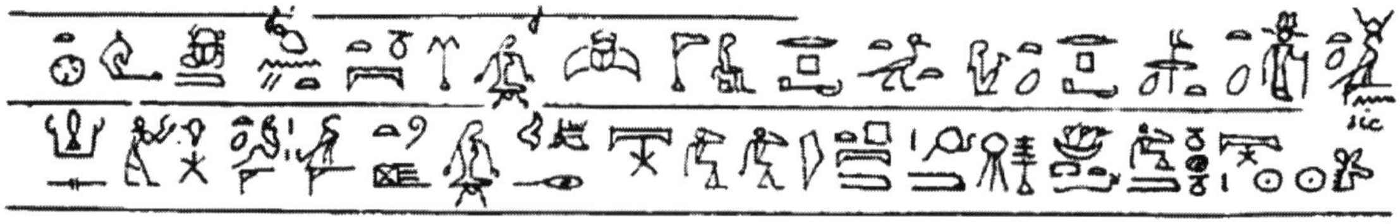

"Horus in female form is the princess, the powerful (*Useret*), the heir to the throne and daughter of an heir to the throne. A flying beetle (she?) *is born in the sky* in the primordial city (Tentyra) *at the time of the night of the child in its cradle*. *The sun shines in the sky in the twilight, when her birth is accomplished*. Gods and goddesses praise the name of her majesty" [east side of the inner wall in the so-called Mamisi, first chamber, of Dendera.]

About the expression so often mentioned in Tententyrian inscriptions: *haru gorh nexen em sestef* "the time (literally: the day) of the *the night of the child in its cradle*", we only know that the intended day *belonged to the 5 leap days* at the end of the year. The more specific information is unfortunately destroyed at the relevant point in the festival calendar list of the temple of Dendera. Since *the 4th leap day is noted as the day of the birth of the goddess Isis*, according to records that have otherwise been received, we may rightly substitute 4 in the illegible position of the calendar. One reads the following:

i.e. "On the 4th leap day of the year that is the beautiful day of the *night of the child in its cradle*, a great festival of the preparations (?). The procession of the goddess Hathor and her venerated gods took place *during the night before this day*. They make a circuit around the temple and carry out everything that is customary. (Then) return to their "places" (i.e. chambers in the temple).

"resplendent rises upon the *golden* (*nubet*, common epithet of Hathor-Isis as the Sothis-Sirius star) over the forehead of her progenitor (i.e., in the immediate vicinity and in front of the sun, hence heliacal), and her mysterious figure is at the apex of his solar ship. If she touches the zenith (*aq*) of her city in the face of her nomos, her dwelling will be viewed in the most joyful mood. When she enters her house, her body is full of delight. When she has taken possession of her noble dwelling, her fellow deities surrounding her, on both sides of her, then the soul in her body is full of joy. When they unite (the fellow deities) with the rays of light of their father (i.e. the sun god) and unite with the splendor of his disk, the city of Anet (Tentyra) is in joy. Praise is offered in *Abet* and *Pi-anet* (other names of the city just mentioned) in a festive mood when they see the Great One, the Sprightly Flowing One, the Creator of Festivals in the Holy City on that beautiful day of the New Year." (Column inscription in the hypaethral temple at the roof of the temple). Right there, downward, you will find the following texts:

"the temple of *Rexit* (special name of the just mentioned temple after an epithet of Hathor-Isis) thrives, in the possession of the *lion* (i.e. of *Ra*, the sun) and *his daughter* (Isis-Sothis), of *Hur-abot* (Horus of the East, i.e. again *Ra*) and the goddess *Xont-abst* (i.e. the one widening at the eastern tip, Isis-Sothis). They take their form in the sky on New Years Day and everyone joins his neighbor" (i.e. Sol and Sirius stand in the immediate vicinity of each other, Sirius rises heliacally.) In another place it is called Hathor-Isis (DBD[29] 46):

[29] Translator note: Dümichen. *Bauurkunde der Tempelanlagen von Dendera.*

"The goddess *Mehennet* (i.e. the diadem) of the god of light (i.e. the sun) and her pilot (*ar-hatef*, actually: the one at the front of her ship) in the solar barque *Sektet*, which traverses the sky always over the head of her father" (a different idea expressing the heliacal Sirius rise). On the north wall in the Pronaos of the temple of Dendera is the name of the same goddess:

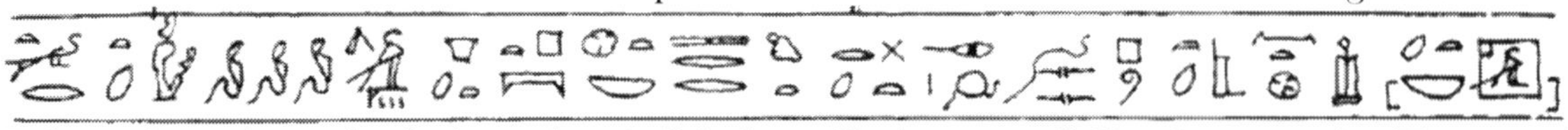

[Hathor, the mistress] of *Anet*, that is Isis herself, the eye of *Ra*, the great in *Tentyra*, the mistress of heaven, the queen of the gods and goddesses, the great *Mat*" (Dikaiosyne) as follows:

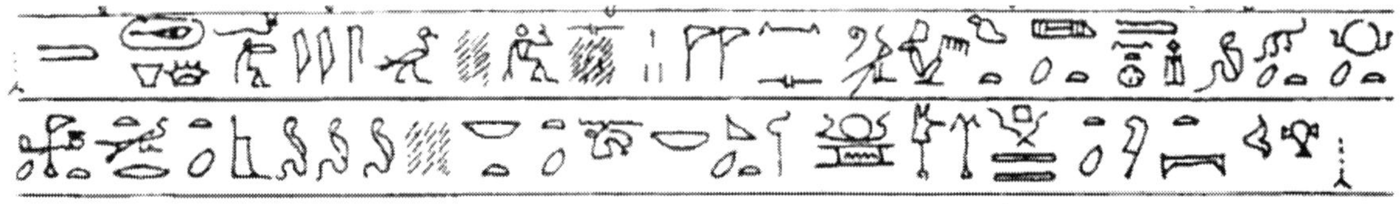

"The feminine sun, the first in Tentyra, the true among the gods ΣΥΝΝΑΟΙΣ [the youthful?] the daughter (?) of a youth, the beauty who appears in the sky, the truth which levels the world at the edge of the sun barge, the queen and mistress of awe, the mistress [the god? and] goddesses, Isis the great, the mother of god." On the north wall of the large hall the same goddess is called:

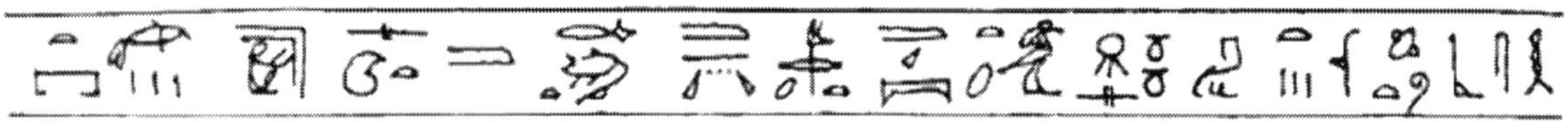

"The years are counted after her rise (i.e. *Sothis Perisden*) she is the radiant one (*Xut*) in the sky, the powerful one (*Useret*) on earth and the very feared one (*a-sonlit*) in the place where the corpses are hidden." On the east side of the terrace wall in Dendera you can read:

"*Uar-xer-ta* (pronunciation uncertain) is the name of this district, *Adut* is called the place of the cradle of Isis. This is usually the house of the birth of the sky goddess *Nut*. The mother of God in the form of a dark-colored woman (or female) [named] *Xnum-anxet*, the mistress of love and the queen of the gods, is born in this place during the night of the child in his cradle.

Her mother said at the sight of her: See (às, ȧs) then I became a mother! Hence the origin of her name *Isis* .... For she is the mistress of the temples of Egypt together with her son *Horus* and together with her brother *Osiris* on this day, from all time to eternity."

According to another version, which is located on a wall (south) of the small temple of Isis west of the great Hathor sanctuary of Dendera, the same text relating to the birth of Isis, i.e. the *Sirius rise*, reads as follows:

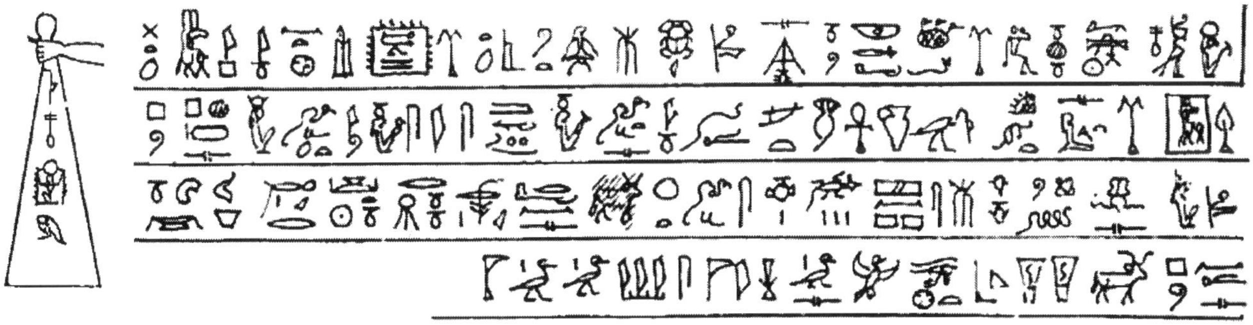

"on this beautiful day (date) of the night of the child in his cradle, on the great feast where the world is brought into equilibrium (*sexex en ta*) the birth of Isis takes place in the interior of *Anet* (Tentyra) by the goddess *Ap* (*Ap*, in the form of a dark red female person, the *Xnum-anx*, the lovely (actually: sweetness of love). Her mother *Nut* spoke at the sight of her: see (ás, ís) I have become a mother! Hence the origin of her name Isis (a passage follows I do not understand). The south towards the rise of the solar disk and the north towards .... She is the mistress of both sides of Egypt with her son [Horus] and her brother Osiris"[30]

On the north wall of the Pronaos, at the very top, following text:

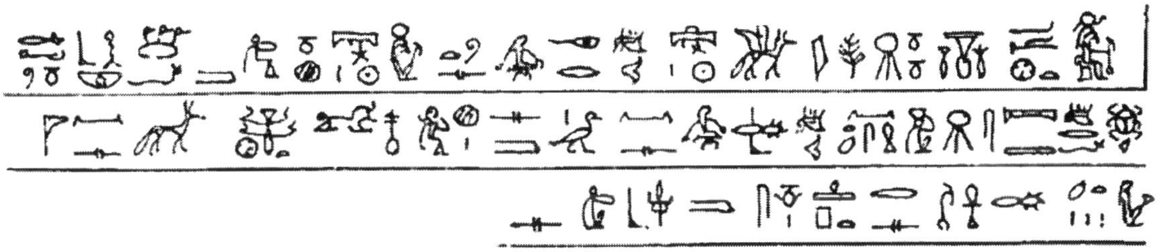

"*Ra-Hur* from Apollinopolis magna, god *Sam-ta*, rises in the twilight (*axex*), when its birth is carried out in the time of the night of the child in its cradle, on the great festival of the whole world (or: the whole country). It shines for Her Majesty when she is born (?when she gave birth?). Your child has the shape of a beautiful boy who is the master of Tentyra. The gods and goddesses bearing the sign of life  and the scepter of power  come to her in order to satisfy her desires according to her wishes."

About the connection between the so-called "*night of the child in his cradle*" (or in his *nest*, in his *bed*, the Egyptian expression *ses* has all these meanings) With the New Year celebrations, inscriptions such as the following give striking evidence:

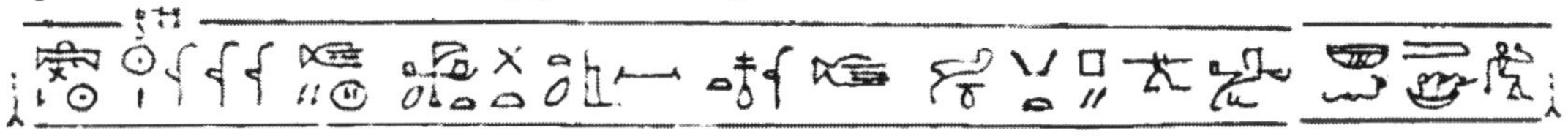

"the bringing of the cloth for the reception of a happy year to the great Isis, the mother of God. [spoken:] Receive, receive happy years on the day of the night of the child in his cradle!"

This text (on the north wall of room A in the temple of Dendera) accompanies the depiction of a Roman emperor who hands the goddess Isis a fabric on which are the words: "Infinite

[30] Translator note: This passage repeats on the previous page in the original addition as well as this translation.

hundreds of thousands of happy years!" as a New Year wish in Egyptian style to the goddess of the temple.

The same scene is found again in the hypethral[31] building on the roof of the temple of Dendera. Here is the hieroglyphic inscription:

"Bringing the cloth to receive a happy year. Text: have a happy year. "

The *heliacal rise* of the Sirius star (on July 20, Julian calendar) formed the starting point for the calculation of the year and New Year's Day. The observation of this rising established a *religious festival,* of which the inscriptions of the temples of Dendera and Edfu were conveyed to us in a thousand-fold repetitions. In solemn procession the statues of the sun god *Ra* and the goddess *Hathor-Isis* (Sothis-Sirius) were erected in their chapels up the stairs ( *Xont* or other synonymous names of the same) from the inside of the temple to the roof of the same (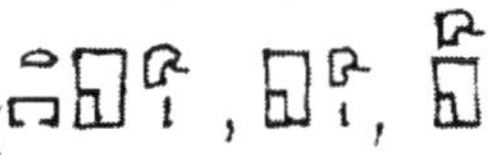 *tep hat,* actually: "Head of the house"), wherever under the open sky or in the hypethral building, in Dendera as:  *hait* most often referred to, the unveiling of the images of God took place at the prescribed moment. The heliacal rising of Sirius was understood as the *connection* of *Isis-Hathor* with *Ra* and as the union of its rays with those of the god of light *Ra,* who at the same time celebrated his birthday as the New Year's sun. On the basis of these indications, one will sufficiently understand the following passages, which I borrowed from Mariette's work *Dendera* (in the 4th volume) for the sake of brief citation. I also want to note that in these and similar texts the Sothis star of Hathor-Isis is referred to as  the "*right eye*", while the *sun* is referred to as  the "*left eye*", different from the usual custom of the holy language in which the Sun as the "*right eye*", the *moon* was understood as the "*left eye*". I must also note that the whole place on the roof of the temples, which was intended for the New Year's celebration, was named 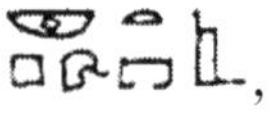, 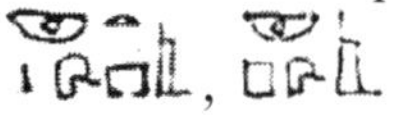 *ust hib tep* "the place of the first festival", often also determined by the image of the hypaethral temple, as in the passage (MD.[32] IV, 1 col. 12): 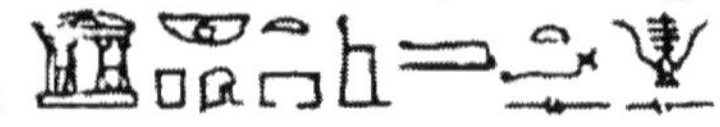 "she (Isis-Sothis) joins her father (the sun) at the place of the first festival" (namely that little open temple). A large number of texts express the idea that Isis-Sothis looks at her father, the sun god, on New Year's morning. 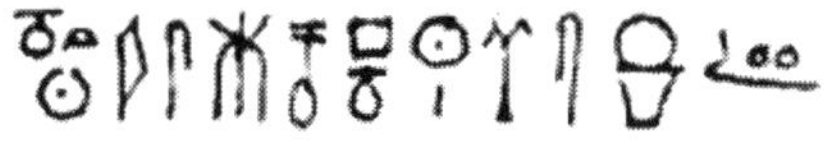 [Isis-Hathor] "looks at her father on that beautiful day of the birth of the sun disc *mas aten,* -- in other places  *mas-ra* "birth of the sun" is used. The goddess is led to the roof  "so

[31] Translator note: In classical architecture, hypethral describes an ancient temple with no roof.
[32] Translator note: Mariette. *Dendera.* Vol. IV

that she can see the rays of her father as he rises" or, in a direct speech to Hathor: 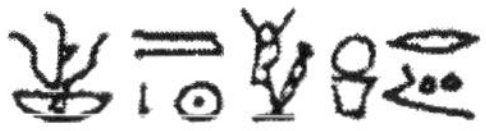 "so that you can see your father on the day of the new year (IV, 20.6.11, etc.). Other texts allude to the Sirius star's proximity to the sun on New Year's Day, such as in following (IV,[33] 3):

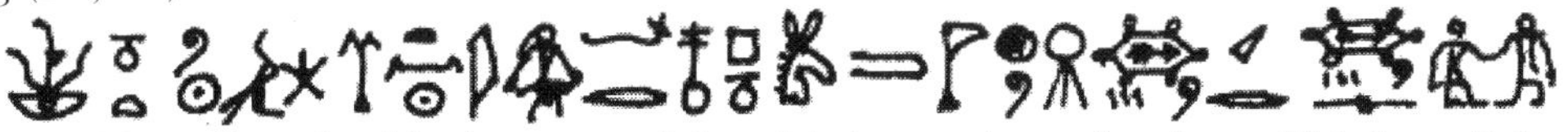

"Its rays combine (*heter*) with the rays of the shining god on that beautiful day of the birth of the sun disk in the early morning of the New Year."

"You join your father Ra (Sol) in your open temple, your beautiful face turned to the *south*" (IV, 2) and others. (IV, 1,12): 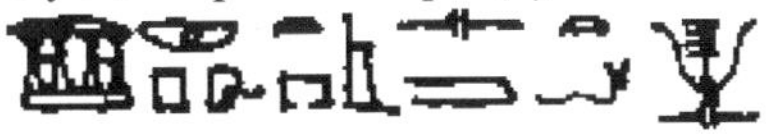 "She joins her Father at the place of the first feast. " Or also as on the same panel IV, 1:

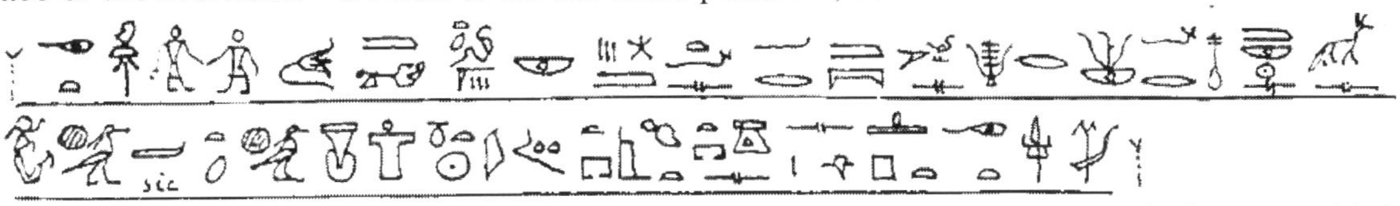

"She comes on her beautiful Feast of the New Year to unite her great ones in heaven with her father, the gods are in a festive mood and the goddesses are full of joy [when] the right eye (Sirius) connects with the left eye (the sun). She rests on her throne in the place where one looks at the solar disk and shining (Isis-Sirius) with the shining (the sun)." Isis Hathor appears

"At their beautiful feasts of the sight of their father, the sky connects with the earth and the right eye (Sirius) unites with the left (sun) *at the beginning of the year, 1st Thoth*" (Dendera).

The eye formula repeats in  ""the right eye, *Isis-Hathor*, unites with the left eye, *Ra* (see MD. IV, 3), whereby the signs do not leave the slightest doubt about the astronomical sense of the deities understood below.

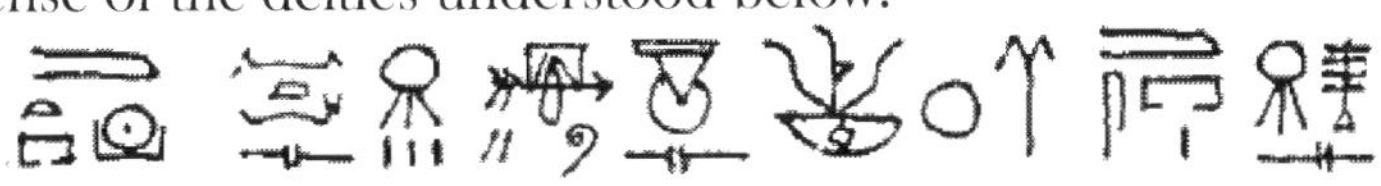

"She shines in her house on New Year's Day and connects with the rays of her father in the sphere of light" (or: in the shining sky) " (Dend. Room E). In this view, the *Tentyritic Isis* is a deity common to all temples and cities of Egypt, hence the lines are cited extensively as:

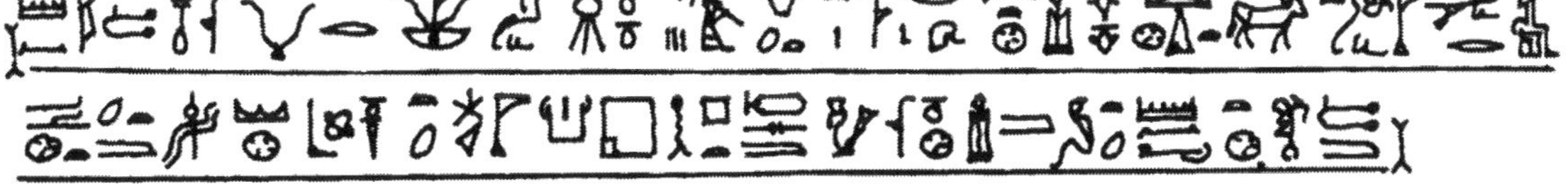

"Isis the great, the mother of God, the mistress of *Adut* in *Anet* (Tentyra), the mistress of the beginning of the year, the woman of the *sema* (?), who *rises on New Year's Day* to begin a happy year, the goddess *Ament* in Thebes, *Manat* in Heliopolis, *Renpit* (that is, the year) in

---

[33] Translator note: Mariette. *Dendera.* Vol. IV.

Memphis, the divine *Sothis* in Elephantine, the brightly shining loin in Apollinopolis magma etc." (Dend. Hall E). According to an ancient concept, it is also the goddess who brings about the threshold of the Nile, which, according to the inscriptions, was tied to the rise of Sirius. Hence the name:

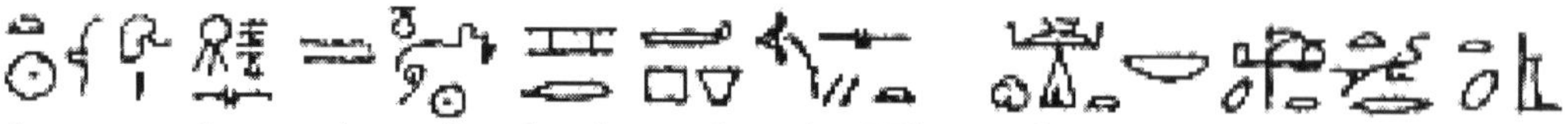

"Isis, the great, the mother of god, who makes the Nile swell at the time when it shines at the beginning of the year" (Dend. Room G). It is therefore also:

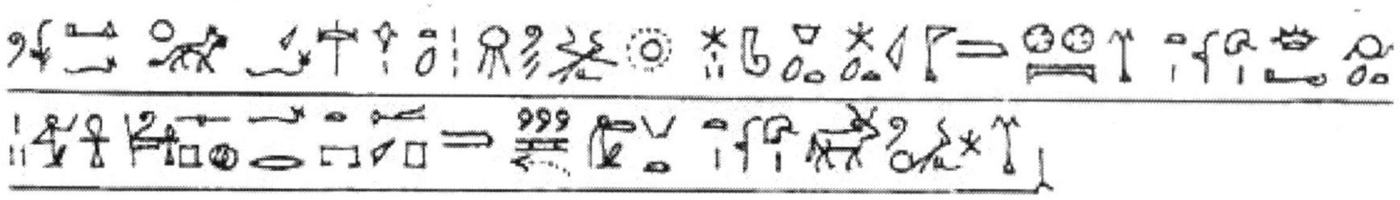

a female sun which *appears in the sky at the beginning of the year* as the divine *Sothis* star, the queen of the decan stars, whose rays illuminate the earth like the sun that appears in the morning, (she is) the mistress of the beginning of the year, who lures the Nile out of its source to bring life to the living people (cf. DHI[34], 49). Likewise, 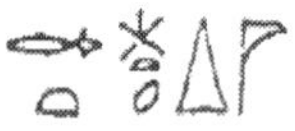 "the divine *Sothis*, the noble" is called in Philae:  "the *mistress of the beginning of the year*, who makes the Nile swell in its time," for which I add the following passages from Dendera that are infinitely easy to understand (see the rich publications of Dumichen and Mariette):  "Sothis in heaven brings the Nile at the beginning of the year to provide food (actually life) for gods and people."

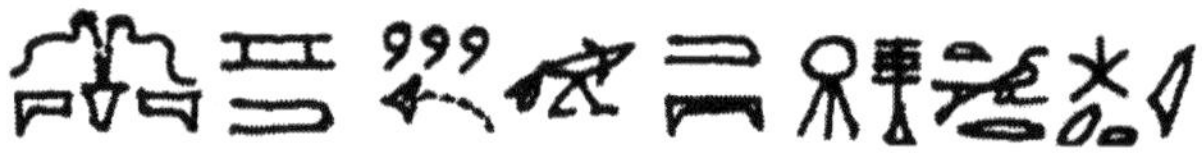 "the great Sothis shines in the sky and the Nile emerges from the two source holes."

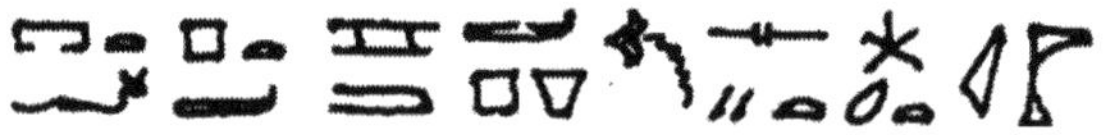 "Sothis makes the Nile swell at its source."

Hathor, it is stated at one point: 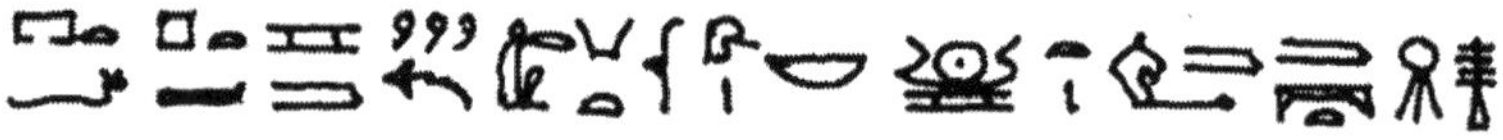
"shines in the sky at the tip of the sun barge and the mistress of the beginning of the year lures the Nile from its source." For this reason, Isis-Sothis is cited as:

 "*Sothis star* in Tentyra, whose *name* is not lacking in any nomos" (Dendera) and thus similar in hundreds of inscriptions.

Like the special *celestial spaces* traversed by the sun, the moons, the planets and other stars and constellations in the hieroglyphic inscriptions as *pir* or *pi* "*house, apartment*," analogous to the Arabic *beth* in the astronomical sense, the same expression is also applied to *Sothis*. Compare the following text relating to *Sothis*:

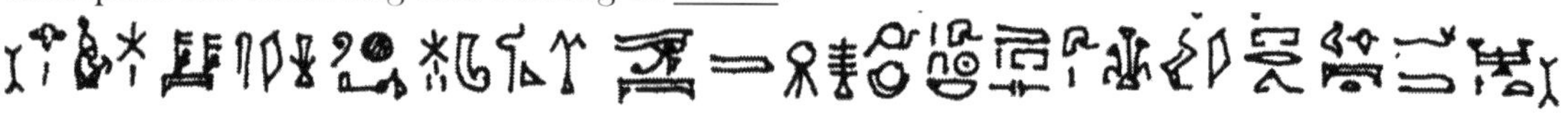

---

[34] Translator note: DHI=Dendera Hall I

"it shines in the sky as the *regent of the decan stars* and protects her brother *Sah-Osiris* (i.e. *Orion*) on his road in the sky, *emerging from their house at the beginning of each decade*, perpetually" (compare DHJ.[35] T.[36] 42, 6 z.[37] 3/4), which is linked with the frequent mention of sacrifices  "at the beginning of each decade," *tep haru met nib*. This is what it says on the stone deed of donation from Philae (with the date of the *Macedonian* month *Peritios* ) that the Ptolemy in question had designated the tithe of the so-called Dodekaschoinos for this purpose:

"To provide the city of Qatar (*gebhut* = Elephantine) with all sorts of good things for his father *Osiris*, the great god, the lord of *A-uabit* (the Abaton) and for his mother *Isis*, the life giver, the mistress of *A-rak* (Philae) for the beginning of each decade" (see ChND.[38] I, 200). Similarly, it says in *Der-el-medineh* (with the temple from Ptolemaic times):

"the great living god *Amenapet* comes to him *at the beginning of each decade* and the Theban god *Xonsu-su* because of their demand(?) of sacrifices."

As is well known - and this fact was first demonstrated by *Lepsius* - the Egyptian year was divided into 36 or 37 ten-day weeks or *decades*, the names of which I have given in detail above on p. 21. They were related to the same number of *decan stars*, each rising at the beginning of a new decade. The first decan star, i.e. at the beginning of the *first week* of the New Year or *on the day of the New Year*, was Sothis or the constellation of Sirius, which rose behind Orion on the designated date *in the normal Sothis year*. So she appears, as it were, as the *regent* of all subsequent decans up to the end of the year. This is what the inscriptions say with great clarity, as one can infer from the following examples.

 "The noble Sothis, the mistress of the *beginning of the year*, the queen and mistress of the *decan constellations* (*Xabsu*), the daughter of *Seb* (of the ag. Kronos), Isis, the great mother of God, the mistress of *Suont* (Syene, Aswan. Text from the Sothis temple near the modern city of Aswan). Similarly, she is called in the texts mentioned above: 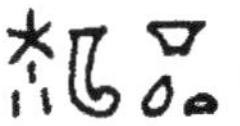 "*Queen of the decans*" and as 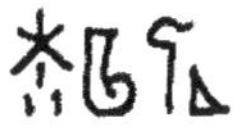 "*regent of the decans*", Isis-Sothis gives the rulers of Egypt the period of their reign, calculated according to the *astronomical Sothis years* (compare the passage cited above  "the years after her rise are paid," compare MD.[39] I, 19, g). Hence the not uncommon formula referring to the Sothis and the reigning pharaoh:

---

[35] Translator note: DHJ=Dendera Hall J

[36] Translator note: T.=Tafel=Plate

[37] Translator note: z.= Zahl; Zeile=number; line.

[38] Translator note: Champollion. *Monuments de l'Égypte et de la Nubie. Notices descriptives*

[39] Translator note: Mariette. *Dendera.*

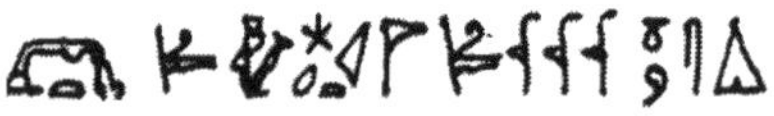 "she gives him *the years of the Sothis of heaven*," or in the version: 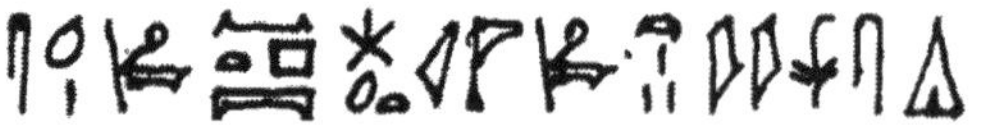 "she gives *the kingdom of the Sothis of heaven* to her son" (i.e. the reigning king). At the *end* of these remarks about the Sothis star, the *mythological* note that the Sothis goddess as the *first daughter* of *Ra*, or whatever name the sun bears as god, tends to appear. In the hypaethral temple on the roof of the Tentyritic sanctuary she is called:

 "Hathor, the great, the mistress of Tentyra, the Eye of *Ra*, the mistress of heaven and the queen of the gods .... who goes up as *Nubet* (the gold), *the daughter of the god of light*, Sothis (the sun), the great, the mistress of the beginning of the year, who makes the river swell to flood the earth. Elsewhere (see MD. IV 25, a):

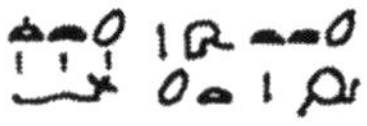 "Ra's daughter, the first of his daughters." There she is called "Pilot of the boat of the morning sun" (*ar-hat-em-sektet*), as is usually the case with "*Diadem*" on her father's forehead (see above) and "*Braut*" (*Sepsit*), such as B. in MD. IV. 24:

 "the temple of Tentyra is *bridal* in possession of *a bride* on her beautiful festival *of the birth of the sun*" (*mas-ra*) and much more all allusions to the (heliacal) *rise of Sirius in the vicinity of the sun*.

The Constellations behind Sothis and Orion

After the constellations of *Orion* and *Isis* and separated from them by the above depictions and names of the three Horus planets *Jupiter*, *Saturn* and *Mars* (see page 54), certain constellations appear in the star tables of the *southern sky* from the ancient pre-Greek epoch in the following order with their relative names:

Constellations behind Sothis and Orion

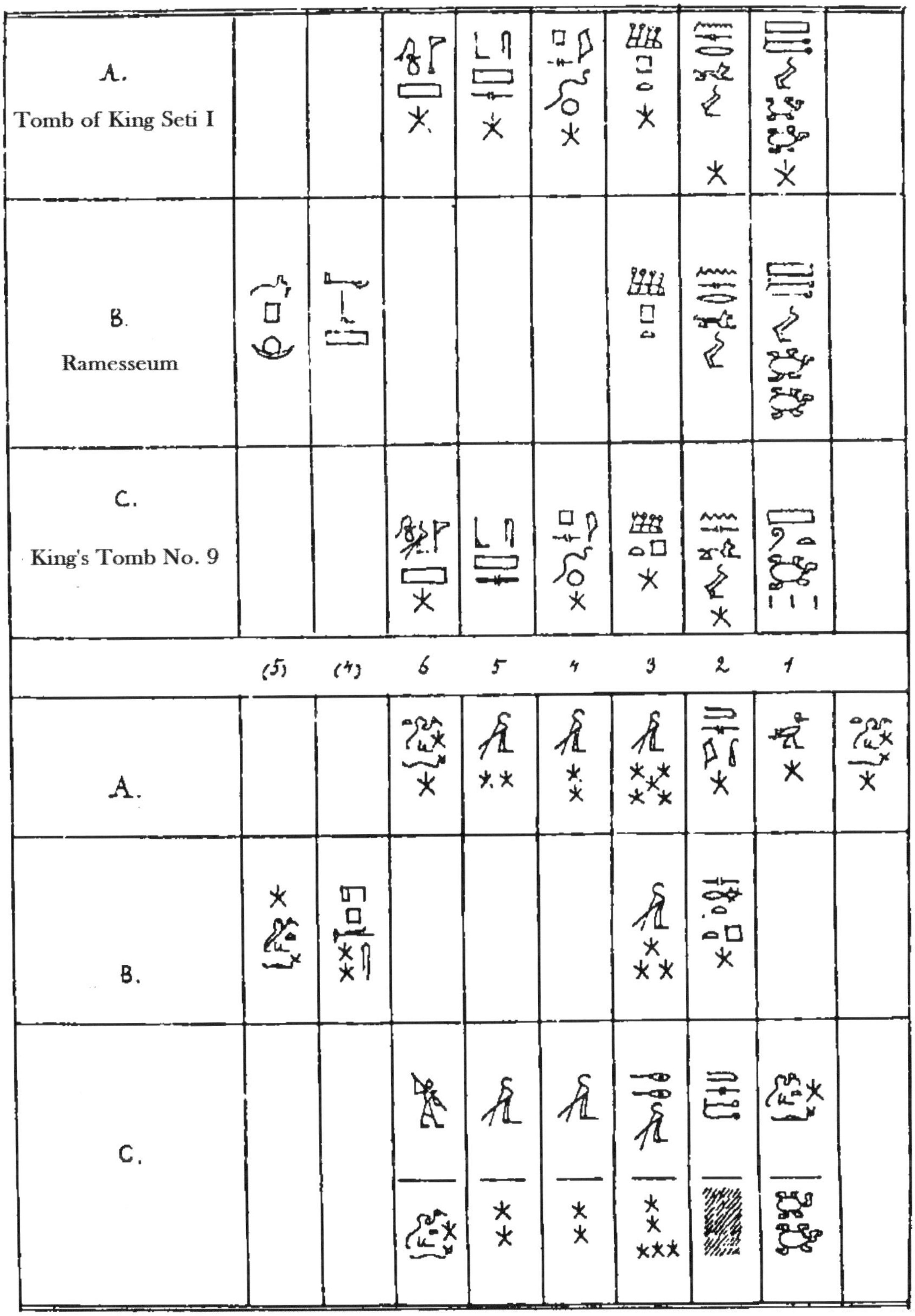

The Constellation of the 2 Turtles

The above panel under groups A.B.C.1 names a constellation šeθu (A.B.) or šetu (C.), which is to be translated with consideration of the following signs: "*the 2 turtles*" or "*the turtles.*" The same is not foreign to the Greco-Roman era, as illustrations and star lists prove. On the sarcophagus of the Theban priest *Hetar*, *the two turtles* appear in front of a seated lion with the head of a hippopotamus or crocodile, 2 swords at his forefeet:

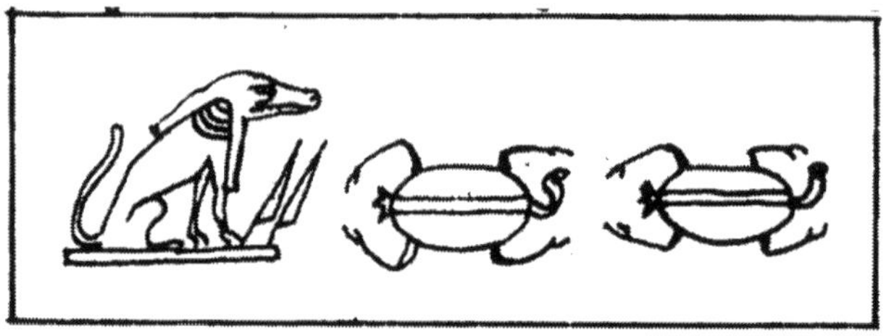

while in the illustration given above on p. 52 (see E) the star of the *turtle* stands between the sun and Sirius.[40] It is consistent with this position that in the more recent decan lists (see below) a constellation , is introduced into the realm of the decans, where it takes its place in front of the decan *Knum* and behind Sothis or Sirius (compare previous p. 18, No. 37). In the Greek list of Salmasius, the Egyptian name šeta, šita or šit is very precisely circumscribed by CIT.

The Constellation *Nestu*

(see p. 85, 2)

Since the word *nestu* or *nesut*, *nesru* lacks the determinative sign at the end; the designation of the same is twofold in the translation. The closest thing to consider is the word *nort*, meaning "flame, fire".

The Constellation of the Sphere

(see p. 85, 3)

Called *sesepet* in the three lists (on the word itself, see my *Wörterbuch* Supp. p. 1131). Otherwise I did not encounter the name.

[40] Translator note: The ancient Egyptians portrayed Cancer as a scarabus beetle, but the Greeks and Romans depicted it as a tortoise or turtle. The sun rose in Cancer in the time associated with the positions of the stars, planets and constellations in the round zodiac of Dendera. See our Appendix B, see also Bunker-Pressler forthcoming book *The New Order of Things*.

The Constellation of the Disk

(p. 85, 4)

Regarding the meaning of the word *apeset*, whose more recent spelling is in the form *apeset*, or *apiset*, I refer to my *Wörterbuch* Supp. 487, see voc. *pesedet*. *apeset* is obviously the *oldest* and *most correct* spelling. For the position of this so-called constellation, which has been omitted from List B (from the Ramesseum), see the remark on the flying constellation.

The Constellation *Seb-ses*

(p .85, 5)

Only mentioned in lists A and C, although it seems doubtful whether in the places concerned it would be better to read , *seb-seta* (that is: "the hidden star").

In the Greco-Roman lists of the decans (see below) a constellation appears: *sebxes* varr. *sebxet*, *sebses* as decan (Greek COYXωC) behind the decan *apeset* (Greek αφοcό). Both take the place of the older decan image: *sebu mahu* i.e. "Full stars," both also appear in King's tomb no. 5 in the form: and *apeset* and *seb-xexet* the latter spelling appears to be erroneous instead of *seb-seth*.

The Constellation *Nutar uas* or *Uas-nutar*

(p. 85, 6)

A name that is frequent in the inscriptions but otherwise undetectable as a constellation, which, like the three previously discussed, does not appear in list B.

The same list B mentions two other names in place of the ignored names, one of which is *abes*, the other *anep*. So much can be stated about the last one with complete certainty that it corresponds most accurately to the designation of the *20th lunar day festival* *hib anep* in the previous list of moons (p. 40).
Assuming that the astronomical ceiling painting in the Ramesseum referred to the renewal of a Sothis period in the reign of King Ramses II (which is chronologically entirely permissible), in which the rising of Sirius took place on the 1st Thoth of the current year of change, the lunar phase *anep* was just to indicate that on this day *the moon was in the 20th place of the lunar month*, offering a new and important material to the calculating chronology.

The hieroglyphic signs below the constellations mentioned above, of which we will encounter more examples later, require a special discussion because of their importance. They consist partly of well-known god names (with the exception of *sexet-pet*, B, 2 and *haptes*, B, (4), which apparently refer to constellations[41]), and partly of one or more added stars, which serve to determine the respective constellations according to the number of their stars. The god names return almost all the time in the (older) star lists, but exclude any idea that the god name only refers to a certain *specific* image, since on the contrary the *same* name of a deity appears to be written with the most *diverse* star names. To explain this striking phenomenon, I have to note that the names in question (with individual deviations in the sequence and designation, which can be adequately explained from the different epochs of their occurrence) *could refer everywhere to the succession of spatial and temporal proportions.* The most instructive example of *spatial proportions* is offered by the Ancient Egyptian so-called *royal cubit of 28 fingers*. Each of the individual finger widths is provided with the *name of a special deity*, the *deities of the star lists* also take their particular positions, as one can see from the following list:

[41] In my opinion there is no doubt that astronomers at the court of *Ramses II* translated the star otherwise known as *apeset*, *apeset*, as *hapetes*, and the constellation *sebseta*, *sebxet* as *sexet-pet*.

The Deities of the 28 Finger Widths of the Royal Cubit

The deities of the 28 finger widths (*tba*) of the royal cubit (*ma suten*) of the ancient Egyptians.

[Compare the illustrations of several ancient Egyptian cubits and their inscriptions in Lepsius' treatise on the ancient Egyptian cubits]

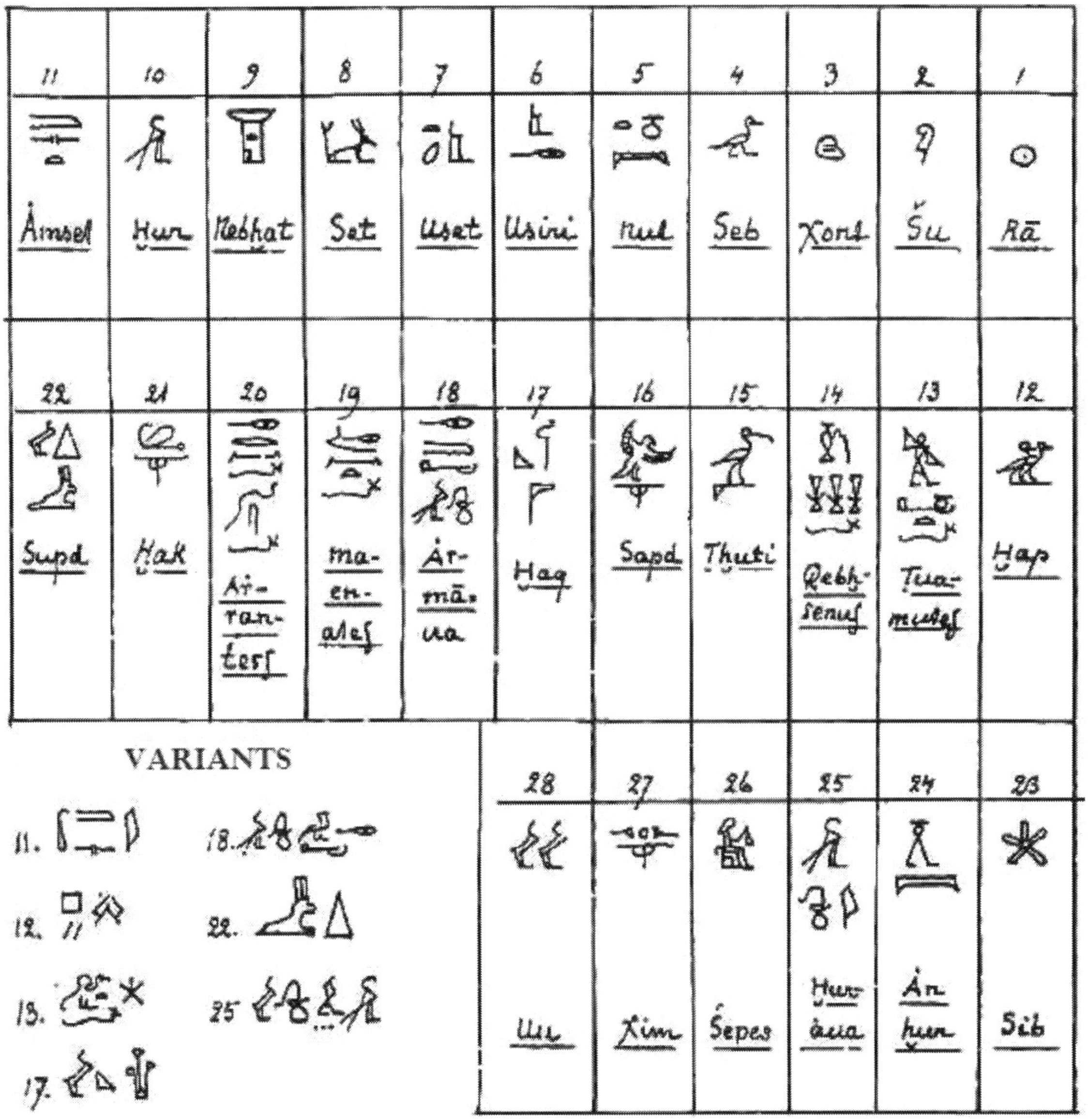

Plutarch teaches us quite extensively about the correlation between *time and space* on the basis of *the number 28*. In his work *Ueber Isis und Osiris* (chap. 42 and 43) he notes according to Egyptian traditions Osiris is said to have lived for *28 years*, and according to others to have ruled. This is the number of days of the lunar month and in that much time he completes his cycle. The maximum height of the Nile at Elephantine is also 28 cubits. The Egyptians pointed to the dismemberment of Osiris into 14 parts (= 28/2) as the days of the waning from the full moon to the new moon. The full moon occurs on the 14th day from the new moon (28/2), but that is also the *cubit measure* of the *middle height of the Nile* near Memphis. Its *lowest* point is at Mendes and Xois; it amounts to *6 cubits*, corresponding to the *crescent* (28/4 i.e. actually 7). To this he adds the remark that the death of Osiris will occur on the 17th, when the setting of the full moon will become apparent. From the above p. 38 ff. with divided lists of the deities of the days of the new moon, it appears that the same deities from 1-9 correspond to the first deities of the cubit (with a shift that must have its specific cause), while in the more calendrical enumeration of the 30 moons (sic) p.39 *Thoth* the new moon, *Horus* the 2nd, *Osiris* the 3rd, *Amset* the 4th, *Hapi* the 5th, *Tuamutef* the 6th, *Qebh-senuf* the 6th, *Ma-atef* the 8th, *Ar-ran-tesef* the 10th, and *Ar-maui* is assigned to the 15th day. The same deities appear again in the Greco-Roman table of hours of the night, but again associated with different numbers, the hours of the night.  *amset* the protection god is in the first hour of the night,  *Hapi* in the second, 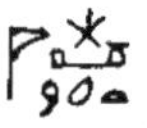 *Tuamutef* in the third,  *Qebh -senuf* in the fourth,  *Haq* in the 5th,  *ar-memai* (sic) in the 6th 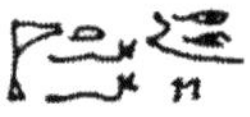 *ma-atef* in the 7th, 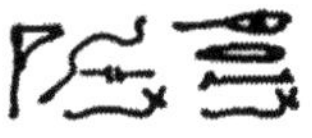 *Ar-ranef-tesef* in the 8th. The connection between the hours and the positions of the moon is given a striking and noteworthy illumination in the inscriptions that accompany a number of divine beings on the walls in the 2nd chamber of the northern Osiris temple on the roof of the temple of Dendera (see MD.[42] IV, 78 ff.). Then appear:

[42] Translator note: Mariette. *Dendera*. Vol. IV

God	Šu in der	1.	Day Hour of the first position of the moon	
God	Hursieusel	2	. . .	2nd . .
Goddess	Isis	3	. . .	3rd . .
Goddess	Soχet	4	. . .	4th . .
Goddess	utit	5	. . .	5th . .
Goddess	menhit	6	. . .	6th . .
Goddess	uer-ḥaket	7	. . .	7th . .
Monkey	āā(n?)	8	. . .	8th . .
Monkey	up	9	. . .	9th . .
Monkey	sa	10	. . .	10th . .
Monkey	Hedet	11	. . .	11th . .
God	mender	12	. . .	12th . .
God	Šu	1	Night Hour .	13th . .
Goddess	[Tafnut]	2	. . .	14th . .
God	Qeb	3	. . .	15th . .
Goddess	[Nut]	4	. . .	16th . .
Female Dog	Anupet	5	. . .	17th . .
Goddess	Xont	6	. . .	18th . .
God	Hur.	7	. . .	19th . .

At this point, unfortunately, the deities depicted in the picture are missing the related inscriptions, according to which the 12th nocturnal hour would have fallen on the 24th lunar day. I notice that in these inscriptions the hours of day and night, as well as the position of the moon, are expressed by their eponymous designations. Those of the lunar days, insofar as they are inherited, can be found in the previous list on p. 39 and following pages under list C".

The remarks that precede suggest that the star-line names of deities do not have any meaning, but contain information related to time in a hidden form. Their precise determination can and will one day serve to provide an essential understanding of the names of the stars. According to the list above on p. 85, the following are mentioned:

1. Constellation Sebu in A Tuamutef and Hapi, in C. Tuamutef
2. Constellation Neslu in A & C. Àmset
3. Constellation Sešpet in A Hur 5 Stars, in B. Same, 5 Stars, in C. Mati-Hur, 5 Stars.
4. Constellation Apeset in A. and C. Hur, 2 Stars.
5. Constellation Sebšes in A. and C. Hur, 2 Stars.
6. Constellation Uaš-nudar, in A. and C. Tuamutef

The name 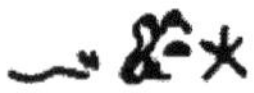 *Tuamutef* occurring in B (5) has such a dubious position in the astronomical image of the *Ramesseum* that it is difficult to determine its relationship.

The Constellations in the Northern Sky

The stars form in the northern sky

In the *Book of the Dead*, XVII, 35, in the well-known stela is:[43]

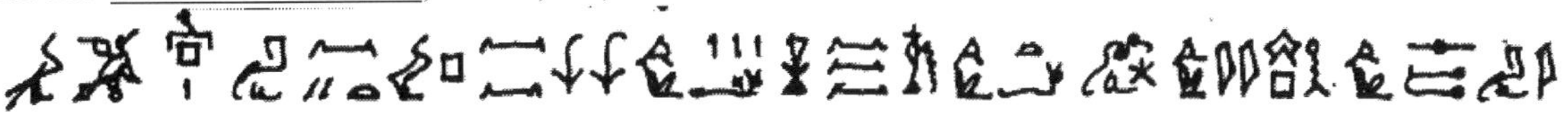

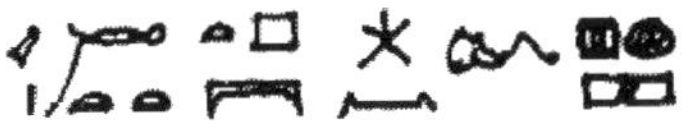 "The gods *Amset*, *Hapi*, *Duamutef* and *Debhsonuf*, these are namely those who are behind the *fore-leg star of the northern sky*," the mention of a constellation: *Xops em pet mahlit* or according to other variations *Xops em pet mahlit* "the foreleg *of* the northern sky" or "*in* the northern sky," already above on p. 67 under the spelling , *mas-Xet* as *northern constellation* par excellence in contrast to , *sah* = *Orion*, the constellation of the southern sky has been discussed. The same meaning of *Xops*, Coptic, rendered as ϣⲱⲡϣ, stella queedam, ⲡⲉ-ϣⲱⲡϣ, and ⲡⲓ-ϣⲱⲃϣ brachinum (this signification is also hieroglyphically verifiable) and *masXet* can be seen from the following passage of the Theban royal tombs, which is repeated several times (G. Ch ND.II[44], 645, 656 ff.):

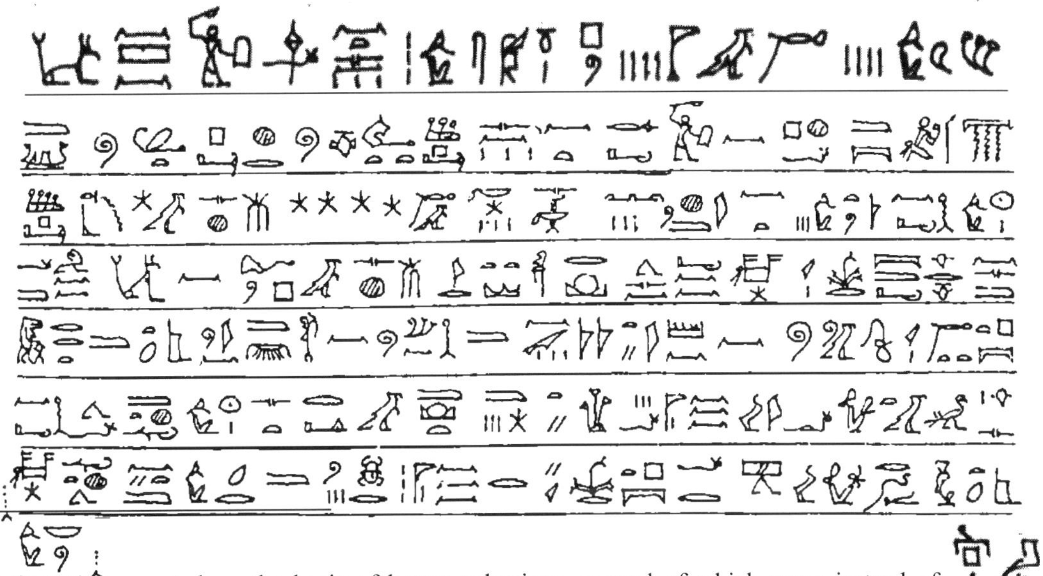

[43] the text is presented on the basis of better redactions, several of which agree instead of *em sa* "behind" *em sau* i.e. read "as guardian" or "as watchman, overseer" (see my Lex. Sup. p. 992). For instead of see line 8 above.

[44] Translator note: Champollion, *Monuments de l'Égypte et de la Nubie. Notices descriptives*.

Translation. "The *four northern* (var. Spirits?) These are the four *gods of the servants* (*ses*). They hold off the battle of the abominable (Typhon) in the sky. He is considered a great fighter. They grasp the front rope and arrange the rear rope on the ship of Ra, in cooperation with the sailors, which are the northern *aXemu-sek*, four constellations (compare above p. 26:). After the *MasXeti*-star ( ), the refuge (compare Coptic) for them is in the middle of the sky on the side south of *Sah-Orion* ( ) and they turn towards the *western horizon*. Concerning the *masXeti-star* ( ), this is *the fore leg* (, *Xops*) of *Set* (Typhon). He is in *the northern sky*. A *rope* is attached to the *two stakes* (*menati*) and to the *knives* ( ) in the form of a bronze chain. It is the office of *Isis* ( ) *in hippopotamus form* (*em reret*) to guard the same" [the text that follows suffers from obscurities that warn against even the most literal translation]. From this important text, to which the following heading belongs: "*the northern spirits*, *four stars* (or constellations) of the sky, the water and the lands of the *northern horizon* (or the n. light side) *above the fore leg*" (hir Xops), undoubtedly highlights the equality of *masXeti* and *Xops*, but at the same time also confirms the passage from Plutarch cited above, p. 66, according to which the star image ἄρκτος for example the great bear, near the Pole Star, to which Typhon was ascribed: the *masXti* star, namely the *fore leg of the set*, it is in *the northern sky*." The usual translation "*club*" of the word *Xops*, however, requires a correction, because from the picture of the celestial cow in the tomb of king Seti I, compared with the inscriptions accompanying the same, all evidence indicates that the group occurring in it *relates to the two forelegs of the cow* and that alone. The names and descriptive designations of stars and constellations contained in the words presented find their most complete confirmation in the images of the earlier and more recent epochs, which appear on individual monuments and *are dedicated to the representations of the constellations near the North Pole*. As you will see, "the front leg, the front thigh" forms the center of the same.

The Constellations near the North Pole according to their Two Main Positions in the Year

1. Representation in the tomb of King Seti at Thebes.

2. Representation from the Ramesseum in Thebes

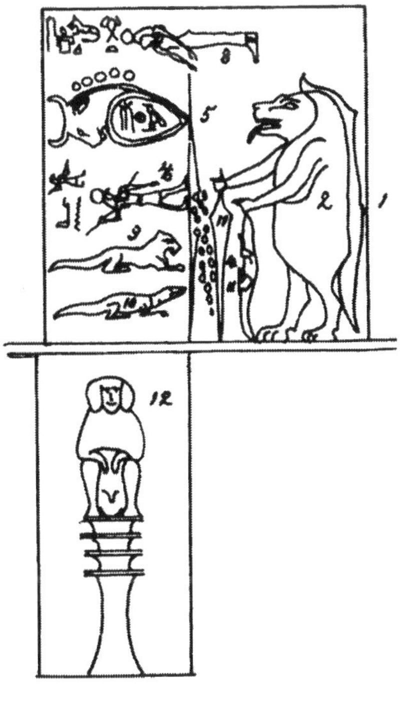

3. Double depiction from a royal tomb at Thebes from the epoch of the twentieth dynasty.

A.

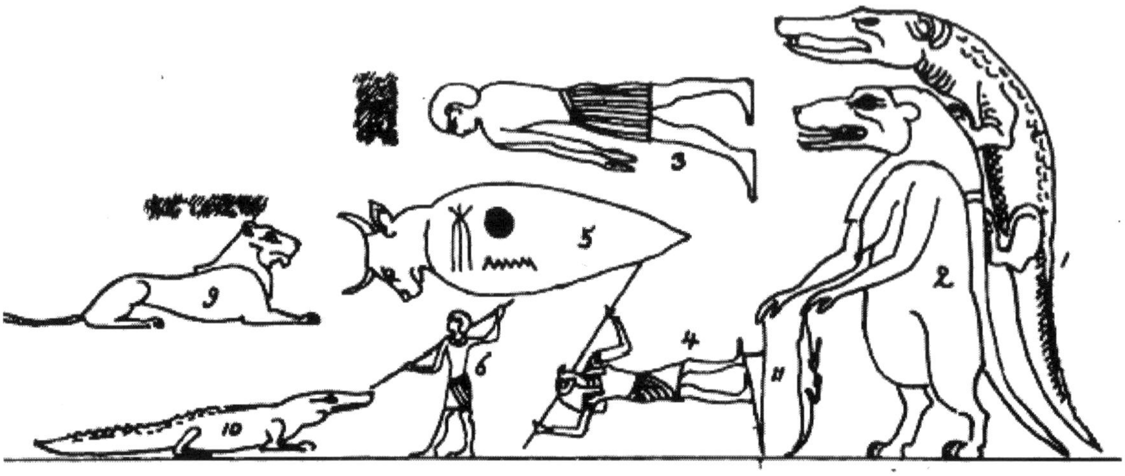

B.

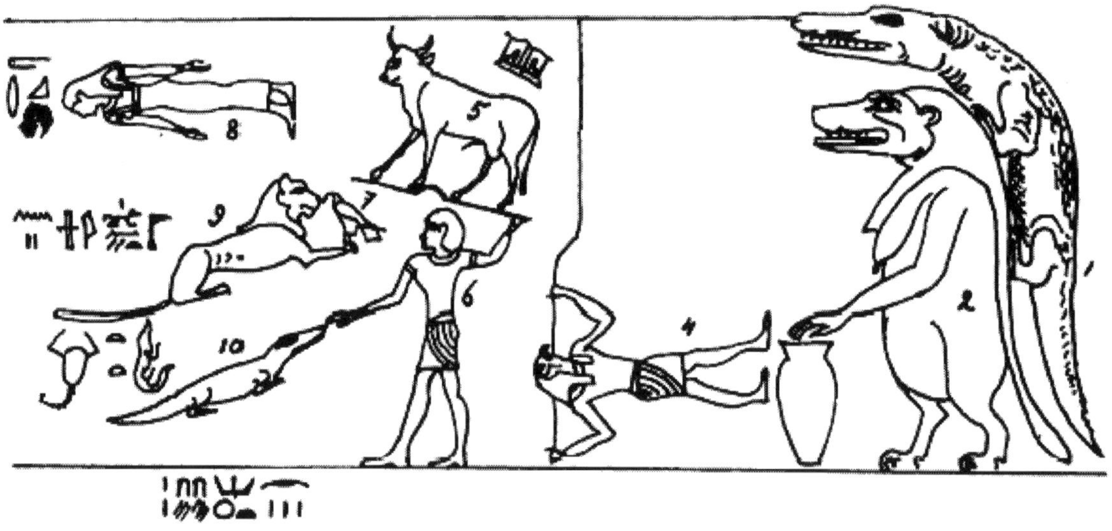

4. Representation from a royal tomb at Thebes from the era of the twentieth dynasty

Representations from the Greco-Roman
epoch of Egyptian history.

5. From the Temple of Philae (Ptolemaic).

6. Depiction from the Pronaos of the Temple of Edfu (Ptolemaic Period)

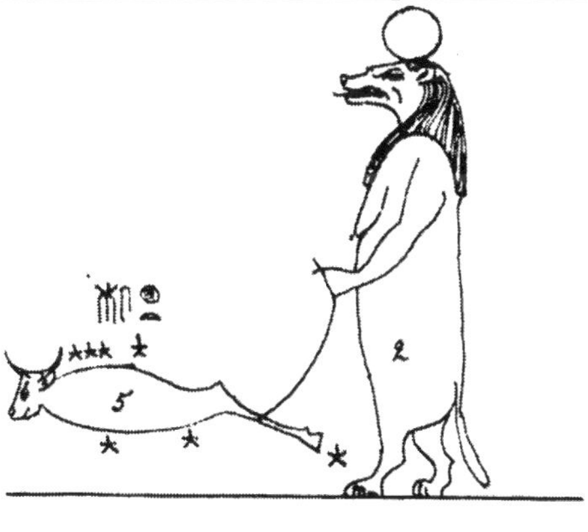

7. Representation from Dendera (Roman time) see above p. 6.

8. Depiction on the coffin of the priest Amon Hetar (Roman epoch).

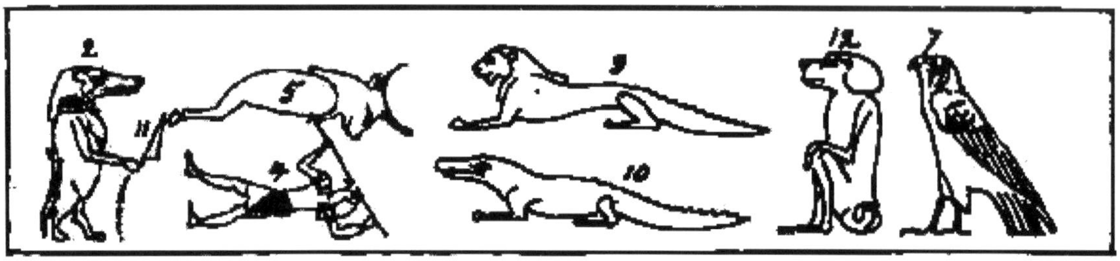

A comparative examination of the individual figures and the accompanying inscriptions, insofar as they are present in the individual representations, gives the following result. I note that attention has been paid to the inscription from the royal tombs given above on p. 93, which undoubtedly relates to the representation in question.

For the sake of brevity, I mark this inscription K.

1. *Crocodile on the back of the hippopotamus*. Without further naming.
2. Hippopotamus, standing upright, Name: *Hesamut* (see no. 7 and 4), in K. On the other hand: "*Isis as a hippopotamus*." Also in the Theban timetables (see below) the same constellation appears under the same name *rert*, with the following listed as individual parts of the female hippopotamus: "*the feet*," "*Her leg*," "*the middle of the knife peg*" (see below no.11), "her _?_", "her *bah* (?)," "her *shame*," " her *chest*," "her *tongue*," "her *two feathers*."
3. Standing male figure, without any closer identification.
4. A Horus, god with a sparrowhawk's head, called *an* "the spiral" or "the turner" (no. 1), also *Xer-an* "*the fighter and the turner*" (no. 2.), in K. *Xer* "*the fighter*," in Dendera more fully called "*Horus the fighter of the enemy*." [see above page 4].
5. Bull or bull foreleg, front. Name *masXeti* (No. 3, A), next to *Xops en Set* "*forelegs of Set*" (K), ἄρκτος, the big bear.
6. Stabbing man, without a specific description.
7. Picture of a standing sparrow without a name.

8. Standing goddess called: 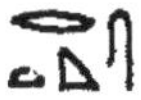 serqet (No. 1) 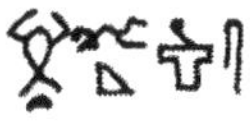 salqet (No. 2),  serq [et] (No. 3, B). The star, if the female figure is to be understood as such, has the same name as a goddess: Serqet or Selqet. The sense that is attached to the word serq is well known. It is called the scorpion. Serqet is "the female scorpion." Also a city,  piserqet,  pi-serket, ψέλχις, ψέλκις, ψελκή, ψελχή bore its name after this animal.

9. Picture of a stretched out resting lion. The inscriptions of the older epochs give it the designation  (No. 1),  (No. 3, B), 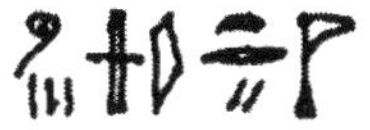 (No. 4), the real meaning of which is obscure to me.

10. Picture of a crocodile, in No. 1 called 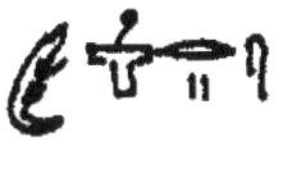 serisa, in No. 3 B and in No. 4 referred to as 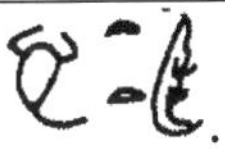. I do not venture to make any assumptions about these names either.[45]

11. In the earlier and later representations of the astronomical monuments, the constellation of the hippopotamus appears in a peculiar way. In No. 1 the left arm is propped up (it is a goddess, to justify the term "arm") on a peculiarly designed instrument , from which a double cord to the rear of the 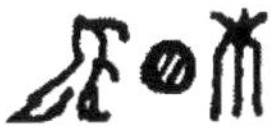 masXeti animal suffices. In No. 2 the right arm has grasped a sword-like instrument [46] with its hand, while the left hand guides an upright crocodile like a walking stick. The small circles  in front of the sword are apparently supposed to represent the rings of a chain. Repeated in No.3 A. the sword, meanwhile, takes the very distinct form: . In No. 3, B. the instrument looks like a large water jug , also in No. 4. In No. 8, from Roman times, but a clear imitation of older ideas, the hippopotamus holds the sword in its hand while a chain seems to be attached to it and hangs down to earth.  No. 5, 6, and 7 clearly show that the hippopotamus holds a rope or chain in his hand, to which the bull's forefoot is attached. In the majority of the examples, the knife and chain are very essential attributes of the goddess of the hippo, and in K. the words allude to them: 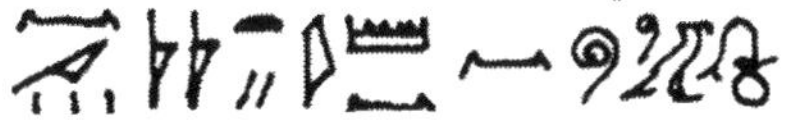 which I have translated above: "a rope is on the two pegs and on the knives," which can also be understood as follows: "a rope is on the pegs (2) of the knives, a rope is on the two knives that are pegged in." The dual  menati presupposes a singular  nebat, which is actually found in the Theban hourly tables in the form  menat. In his learned treatise on the Theban Hours[47] (1871), Dr. Gensler assigned the meaning of "singer [a

---

[45] Translator note: For more information see Bunker-Pressler, *Secrets in the Stars of Ancient Egypt.*

[46] Translator note: it is the pivot of a set of scales for balance, i.e. equinox.

[47] Translator note: Gensler. *Die Thebanischen Tafeln stündlicher Sternaufgänge*

female mammal]" to the word, but gave it a second different translation in the connection:  "*between the thighs* of the [female] hippopotamus" instead of "the *middle of the upright knife* of the hippopotamus." The 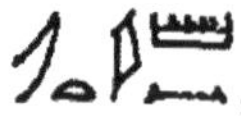, 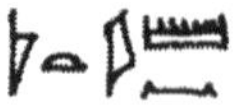 *menat* forms part of the great constellation  *reret,* the hippopotamus, whose individual stars are referred to as: 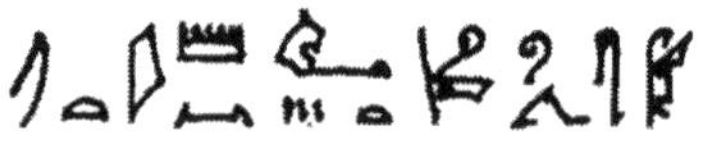 "*the servants* (or, what is more likely, the puller," see my *Wörterbuch,* p. 1407, 2, and compare the north stars of the 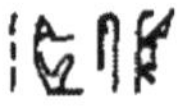 *sesu* mentioned in K.) The *tips of the menat,* then 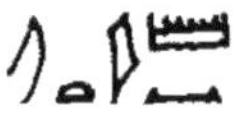 *menat,* 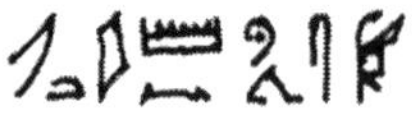 "*the puller of the menat,*" the "*feet of the hippopotamus,*" "her leg," and then the previously mentioned:  "*the middle the menat.*"

Note. Recalling my *Wörterbuch* Sup. p. 600 ff., I also do not dispute the possibility that the word 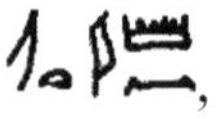, 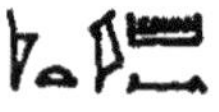 may have the meaning of *support* (of the knife), i. e. the knife's attachment, the knife's handle or the knife's edge.

The Constellations of the Decans

What *Champollion* first wondered about the name and meaning of the so-called decan star images, *Lepsius* later demonstrated in the most striking manner. The 5 decan lists from ancient and more recent eras of the Egyptians compiled by the old master of Egyptologists in the "*Introduction*"[48] on pp. 68 and 69, compared to the *Greek* paraphrases in the list of *Hephaestion*, leave no doubt about it. The following tables contain (with some corrections in the writing and reading of the names) the same lists, but these have been augmented by 7 new directories apart from 2 double lists in individual royal tombs at Thebes. The full number of lists available to me is thus *fourteen*. I will list of them here.

A. *List from the gold chamber in the tomb of Seti I* (19th Dyn.) according to *Lepsius*.
B. *List from the so-called Ramesseum from the time of Ramses II*, based on my own copy.
C. *List from the tomb Ramses IV* (20th Dyn.) according to my own copy.
D. E. and F. *Lists from the royal tombs* (20th Dyn.), amongst them, not included, *two double edits* (C and D). My own copy.
G. *List inside a sarcophagus in the Museum of Berlin*, with the name of King Nektanebos (30th Dyn.) according to *Lepsius*.
H. *List from the pronaos of the temple of Edfu* (Greek epoch) according to my own copy.
I. *List from the pronaos of the Temple of Dendera* (Roman epoch) strip B' according to my own copy.
K. *List from room XII of Dendera*, by *Mariette* and *Dumichen* publication.
L. *List from the pronaos of the same temple*, strip A'.
M. *Directory on the round picture of Dendera*, compared with the paper impression of the original stone set up in Paris in the Berlin Museum.

A closer examination of the lists, and a comparison of them among themselves, first establish the fact that the *older lists from pre-Greek times are based on a different scheme* than the lists of the decans from *the Greco-Roman period*. The existing difference is also shown in another way, in the fact that in older lists the names of the decans are accompanied by those deities discussed above p. 89, which relate to *spatial or temporal succession*, whereas in *the Greco-Roman period* the forms of the divine companions were entirely *different types* of the above-mentioned offerings, and indicate a later origin that can be traced back to the times of King *Nectanebo*. Another remark concerns the *coherence* of decans standing next to one another, which appear in the two oldest registers A and B *arranged according to groups*, without it being possible to give any details about the principle of this arrangement. On the following tables I tried to make the groups visible to the eye by using *double lines*.

The decan stars, in their entirety, appear under different names. The most common and general (in terms of sense) designation of the same is 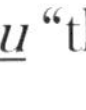 *sibu* "*stars*" (compare p. 67 above) or as in the Ramesseum (p.70) *sibu sepesu* "the *splendid stars*." Another is ,

[48] Translator note: This probably refers to *Die Chronologie der Aegypter*.

anXu "the rising" (stars) or also "biu anXu "the souls of the rising" (see p. 12, line 1). In other texts they are called: sau "the protective stars," or nutari, the divine ones" (see the examples below). Again in the inscriptions they are designated as bekti, bekt, bekti-stars, with uncertain meaning of the word bekt or beket, boket. Its special name was , , , , and many similar variants (compare pp. 64, 68, and my Worterb.[49], p. 1031) with the meaning of: "Lamps, lights" compared to , , Xebs, Xabs, kopt. ϧНВС lucerna). At the head of all the decans was Isis-Sothis, who therefore uses the names: "The Sothis star as the regent of the decans" (see page 64) "Sothis ... the queen of the decans (page 12 line1). In the normal Sothis year, Sothis rose in the eleventh hour (or in the morning) of New Year's Day (compare above p. 72), which is why it is called "Sothis, the great, the mistress of the beginning of the year" or New Year's Day (see p. 83). The successive rises of the decans were noted at the beginning, i.e. on the first day of the ten-day weeks or decades (tep haru met neb), hence the mention of "rise of the stars (i.e. the decans) at the beginning of each decade" (see above, p. 70, B').

In the course of 1416 years of transition, each of the decans had to rise once on New Years Day during a four-year period. In this case his name is neb tep renpit "Lord of the beginning of the year" and appears as tep "head, top" age following decans. Hence the talk of "the decans behind (or: near) them (i.e. of the sun and the moons). The (each time) lord of the beginning of the year counts as the first for them" (compare above p. 68 [50]).

---

[49] Translator note: *Worterbuch*; i.e., dictionary.

[50] Translator note: see B.1, "In the courtyard of the temple of Edfu."

Inscriptions Relating to Decans and Decades

A. Inscriptions from Ombos (Ptolemy era).

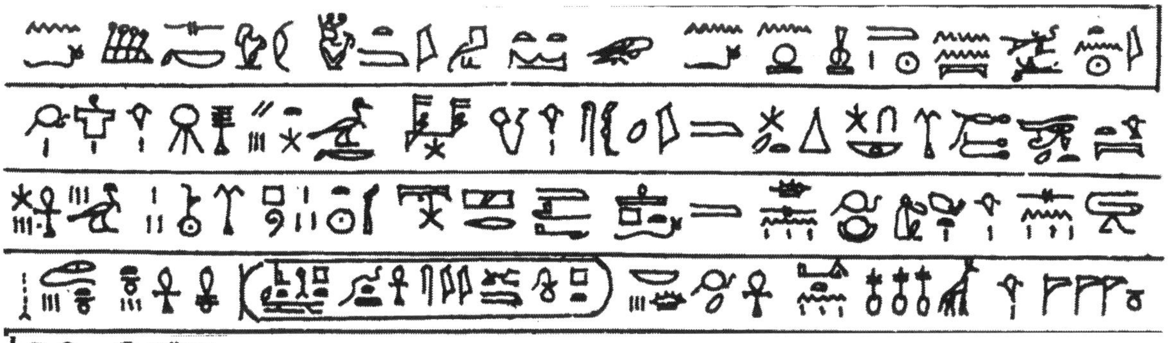

"The large sun disk, wandering through the sky during the day, has covered the orbit on the western side as the god *Atum* (evening sun), then the moon takes possession of the sky. The full moon occurs on the 15th day of the lunar month. (*Isis*-)Sothis in her form (or: according to her manner) faced the *Sah*-Orion star. The *decans* radiate toward the sun. They run in circles, constantly taking turns (*adon*), they appear at their (the sun's) setting in the evening, namely in the hours depending on the seasons. O you souls of the *rising stars* of the gods for the promise of the home, let the son of the sun, the lord of the diadems (*Ptolemy*, the ever-living, friend of Ptah and Isis) ascend, just as you ascend. Protect him from all hardships!

"Regarding the translation "upwards" of  *anx* it should be noted that the double meaning of the same, *"live"* next to *"ascend"* cannot be reproduced by the German translation.

B. Inscription from Dendera (Roman times).

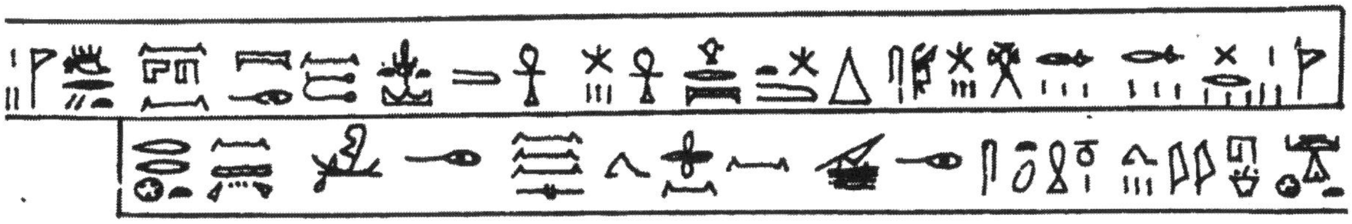

"The sublime and great and oversized gods, the *protective* stars (decans), who follow *Sothis* in the sky, *the ascending stars* (decans), which rise in the east of the sky, who give their care to the deities of *Tentyra*, the messengers. Her Majesty [i.e. of Isis-Hathor] who destroys him who crosses her waters (i.e. acts against her will) and give her protection to the city of Tentyra."

C. Inscription from Ense (Roman times).

"... .. just like the *kektipn*(?) gods on this day of the month, on this month of this year at all their hours. The image of the (decan) *Knum* is set up and the first of the *divine stars* (the decans) is ready in their orbit, the *Bekti* stars, the *ascending* (*anX*) do their duty in the twilight. Star (Orion) and the divine *Sothis* (Sirius) goes down. The great god becomes young (?) in ....

I. The Decan Lists of the Older Period

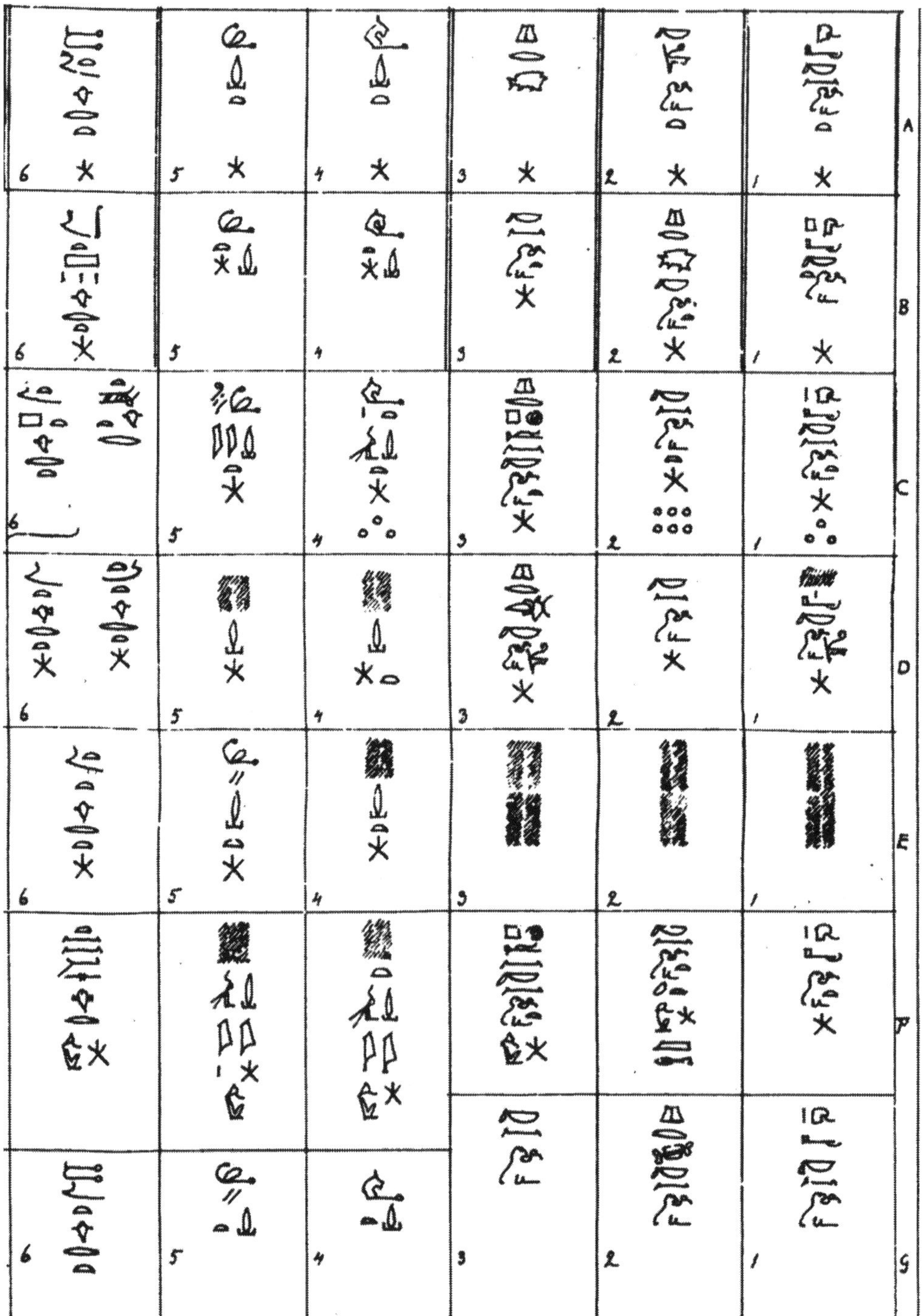

The Decan Lists of the Older Period [continued]

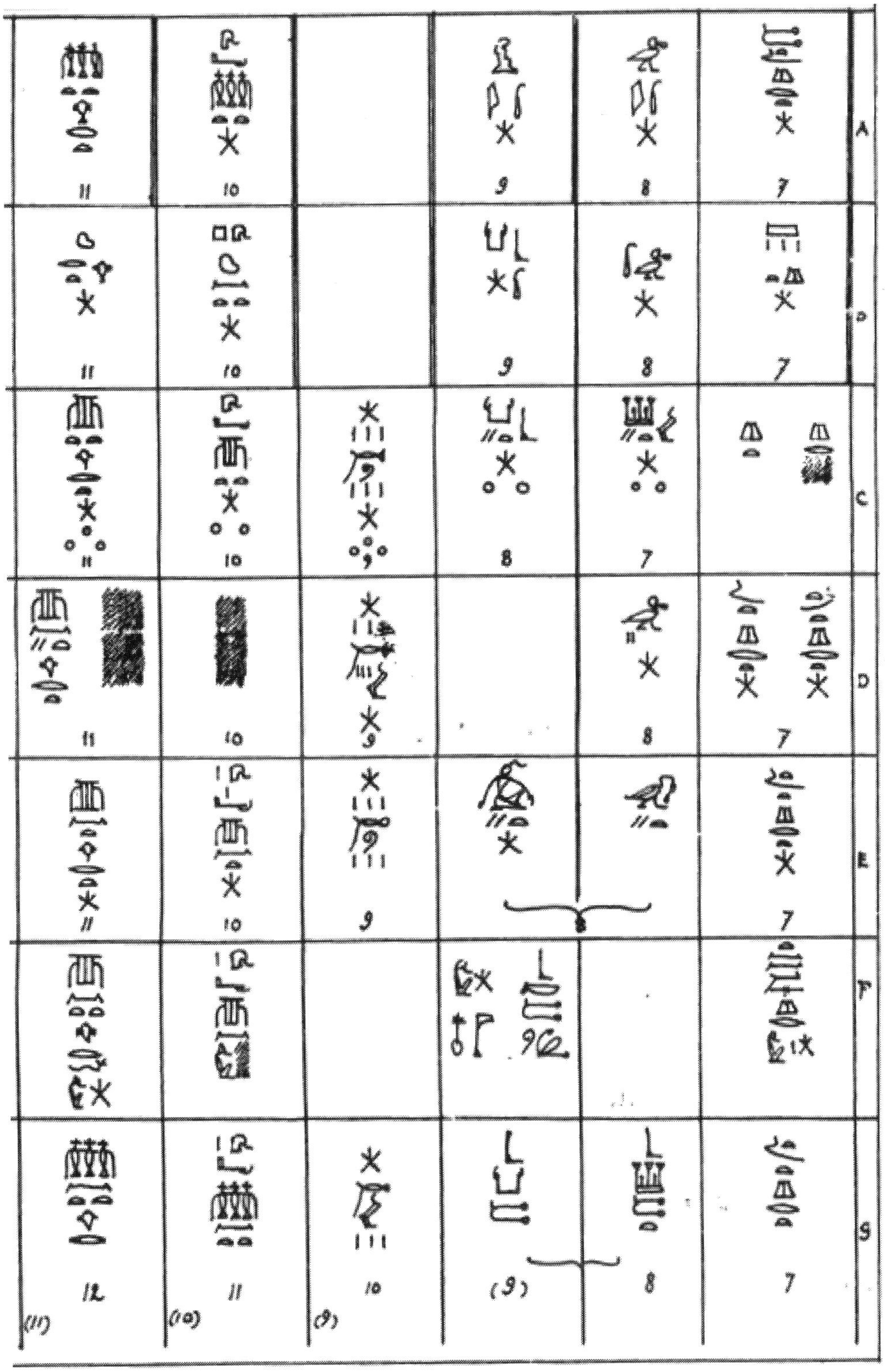

The Decan Lists of the Older Period [continued]

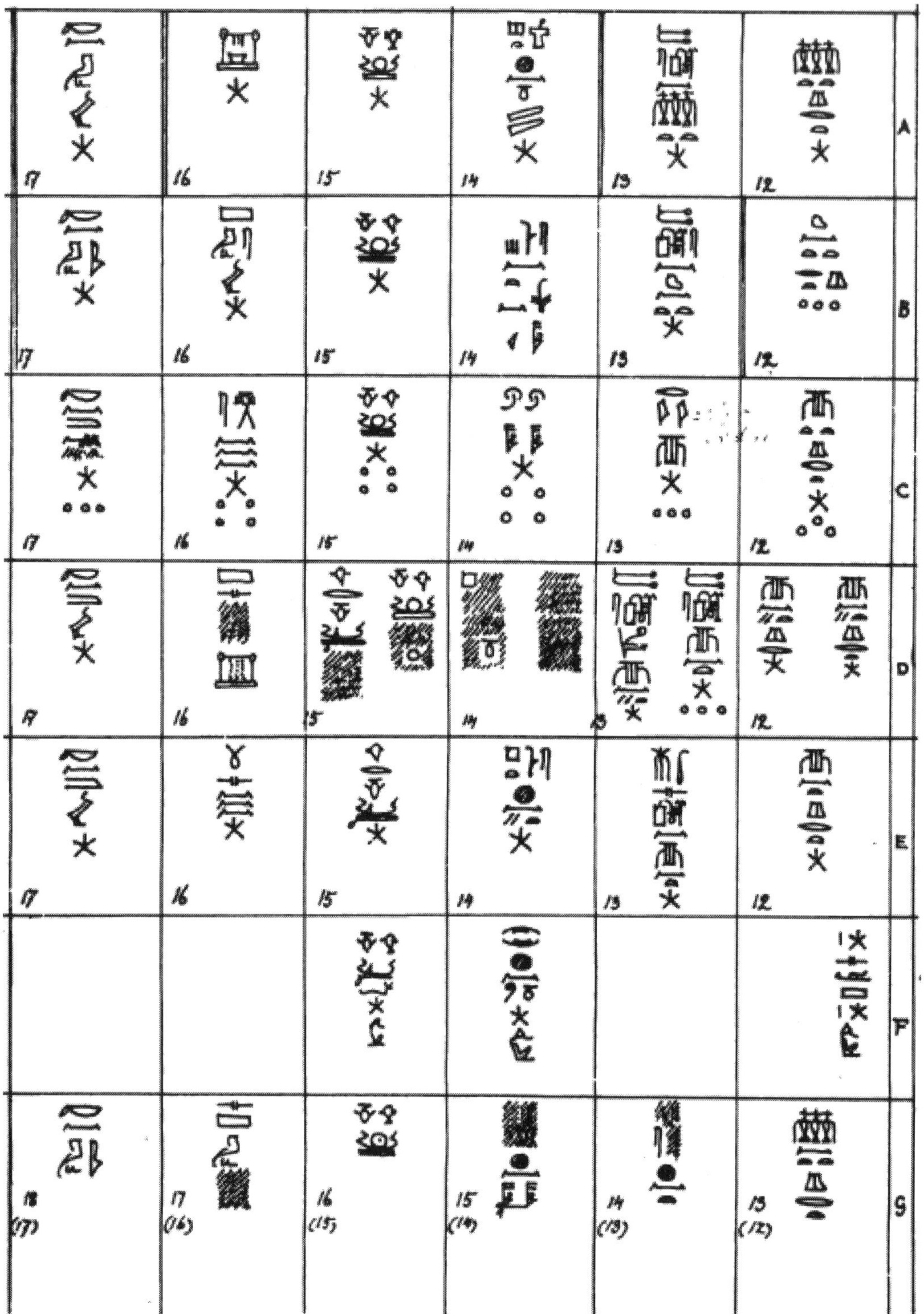

The Decan Lists of the Older Period [continued]

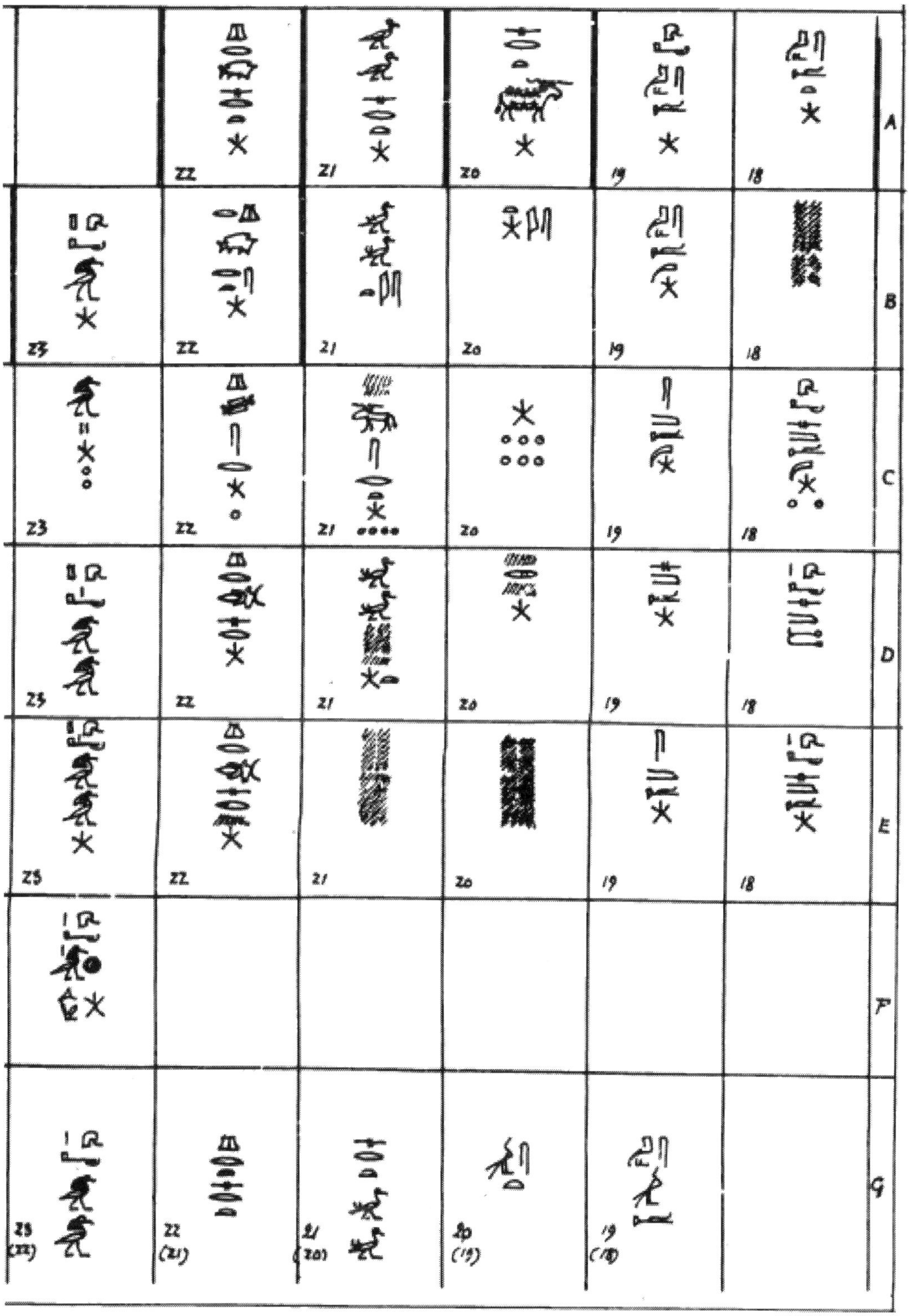

The Decan Lists of the Older Period [continued]

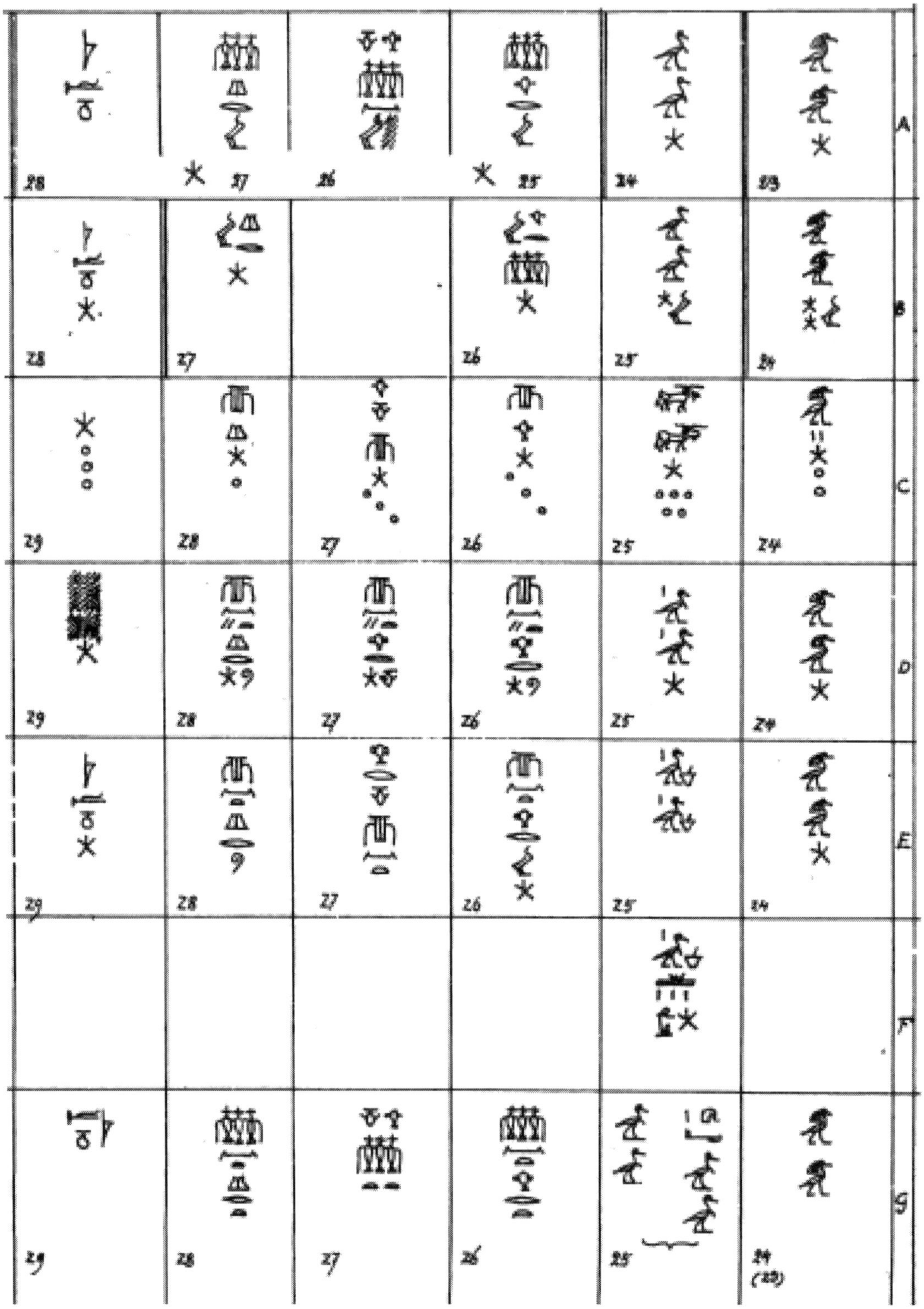

The Decan Lists of the Older Period [continued]

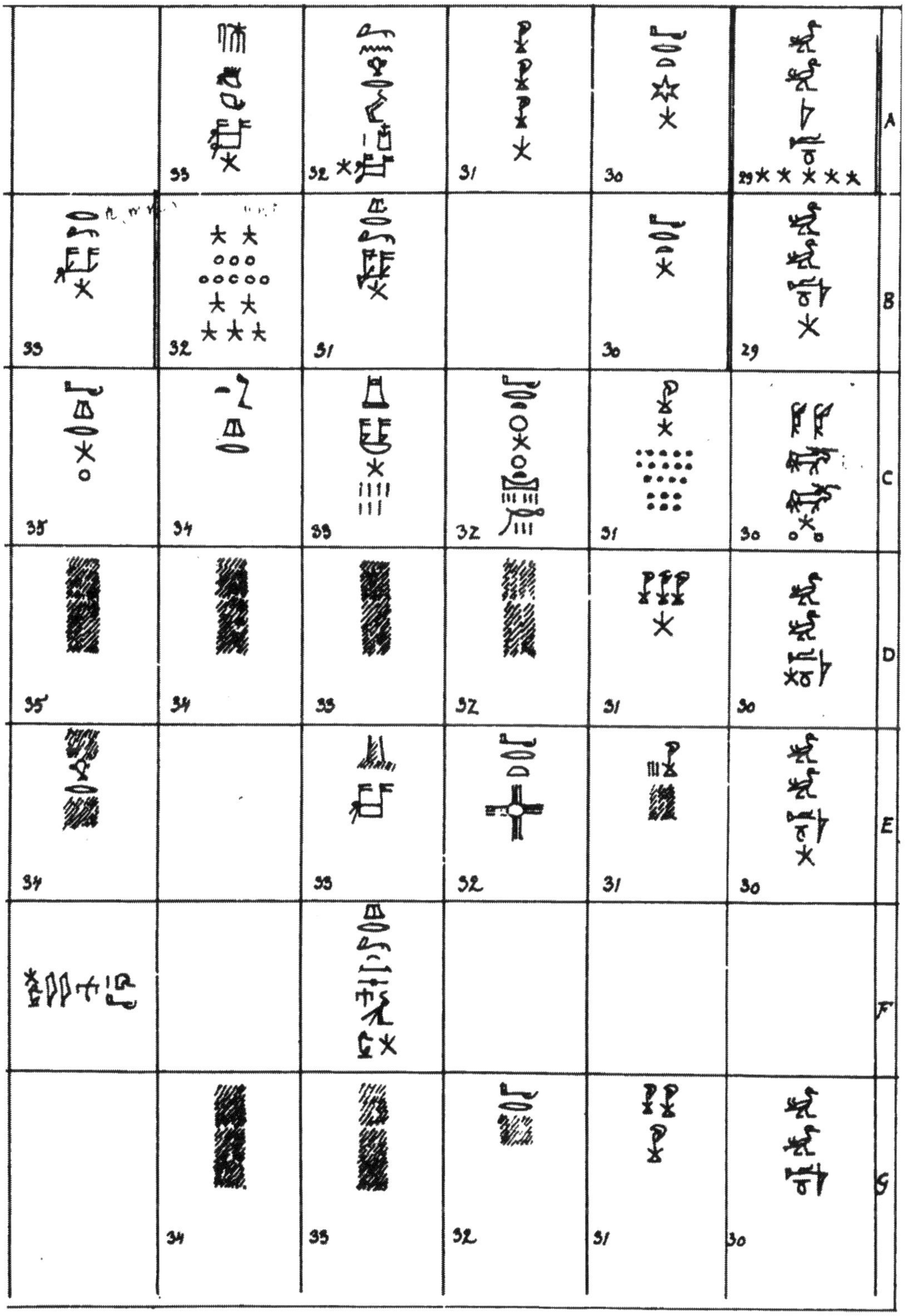

The Decan Lists of the Older Period [end]

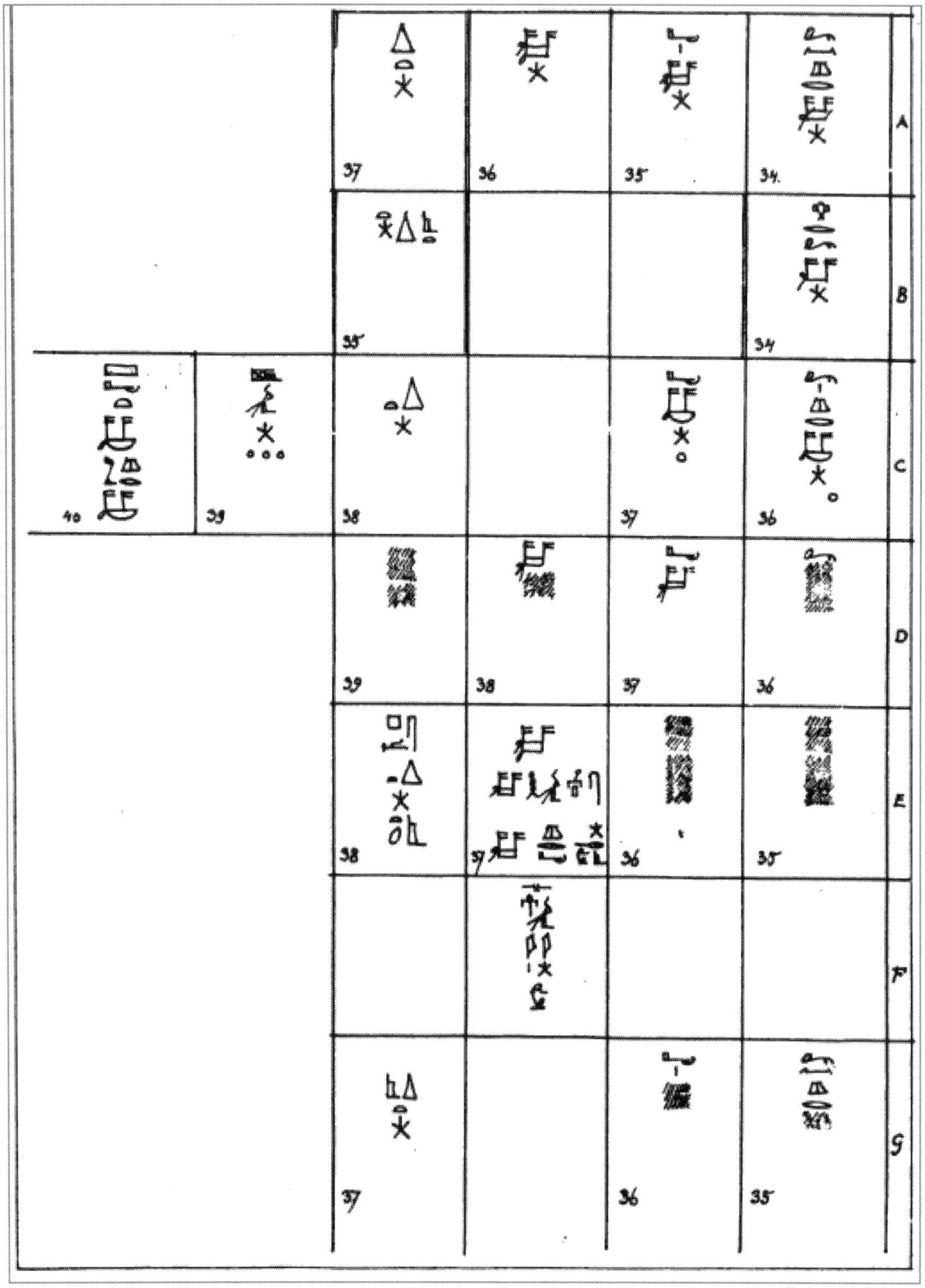

The Deities of the Decans in the Astronomical Sense

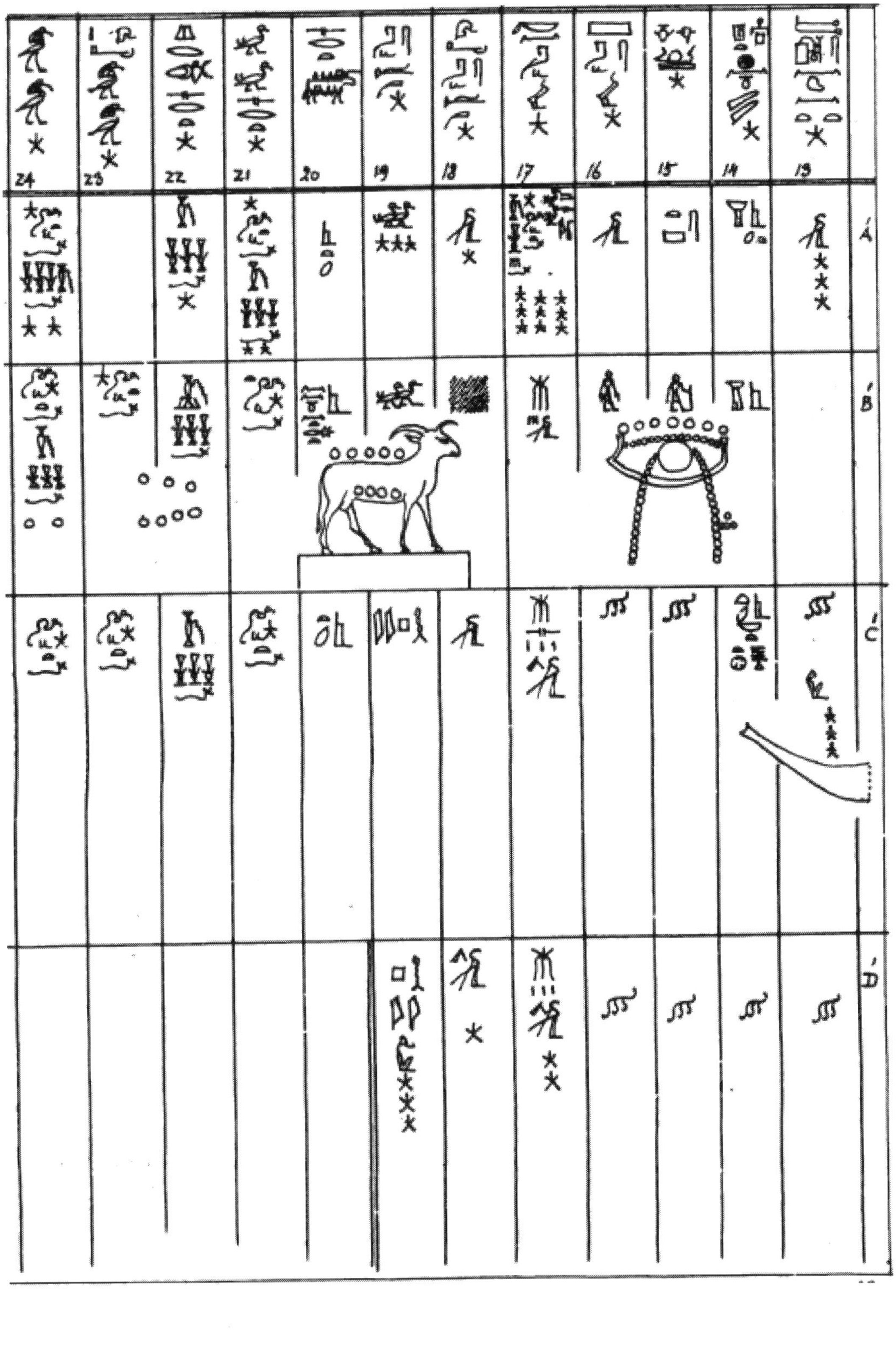

The deities of the decans in the astronomical sense [continued]

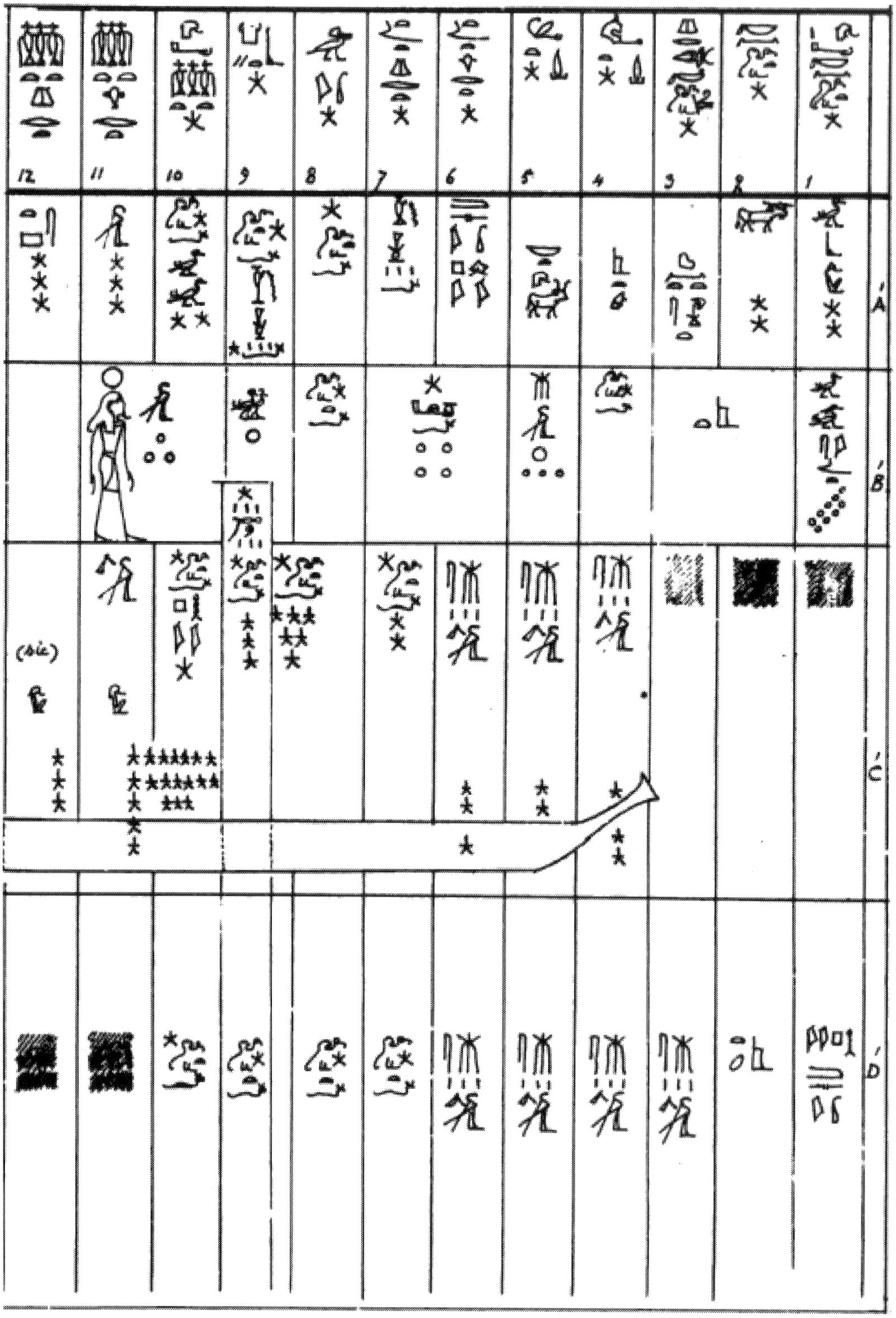

The deities of the decans in the astronomical sense [continued]

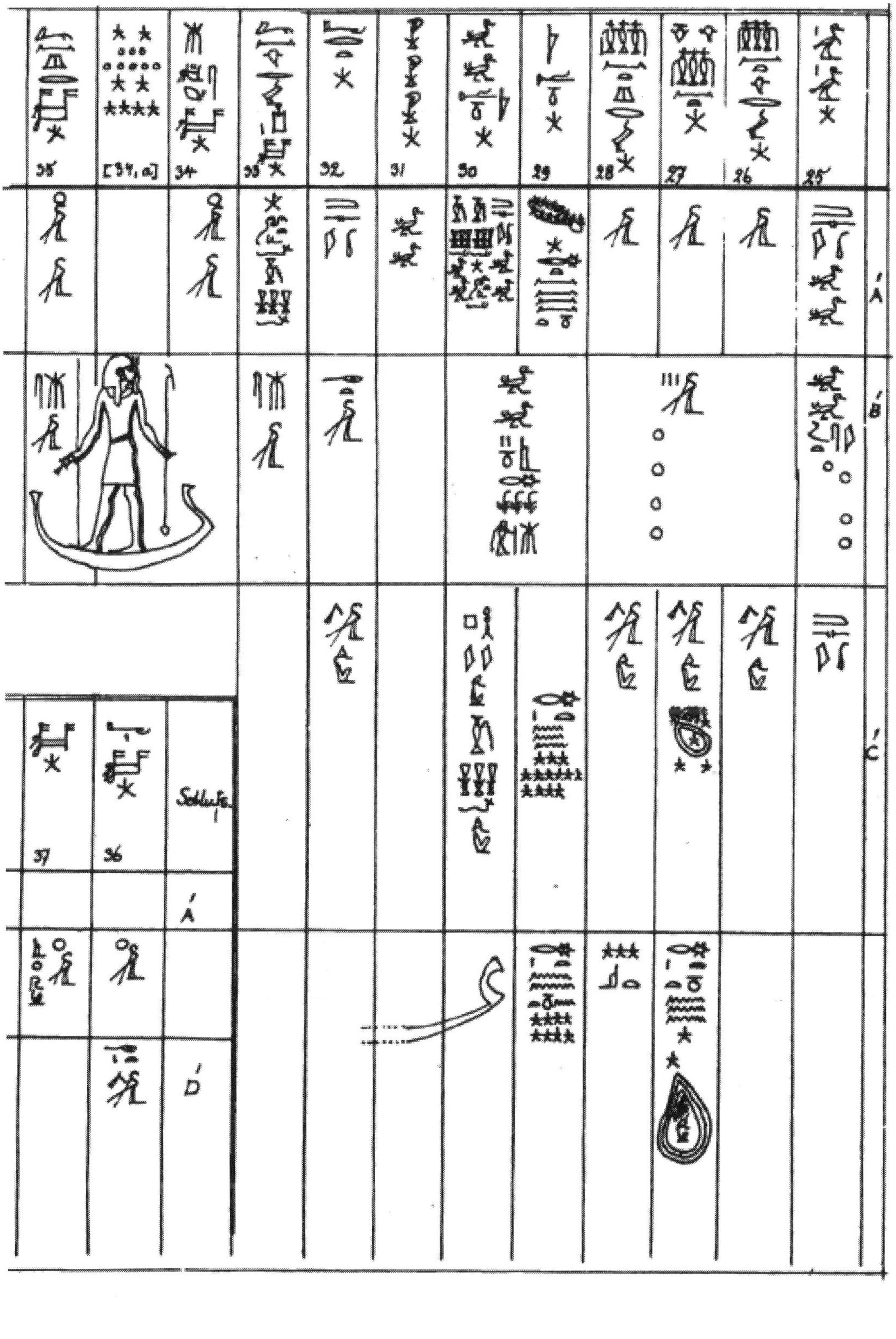

II. The Decan Lists of the Greco Roman period

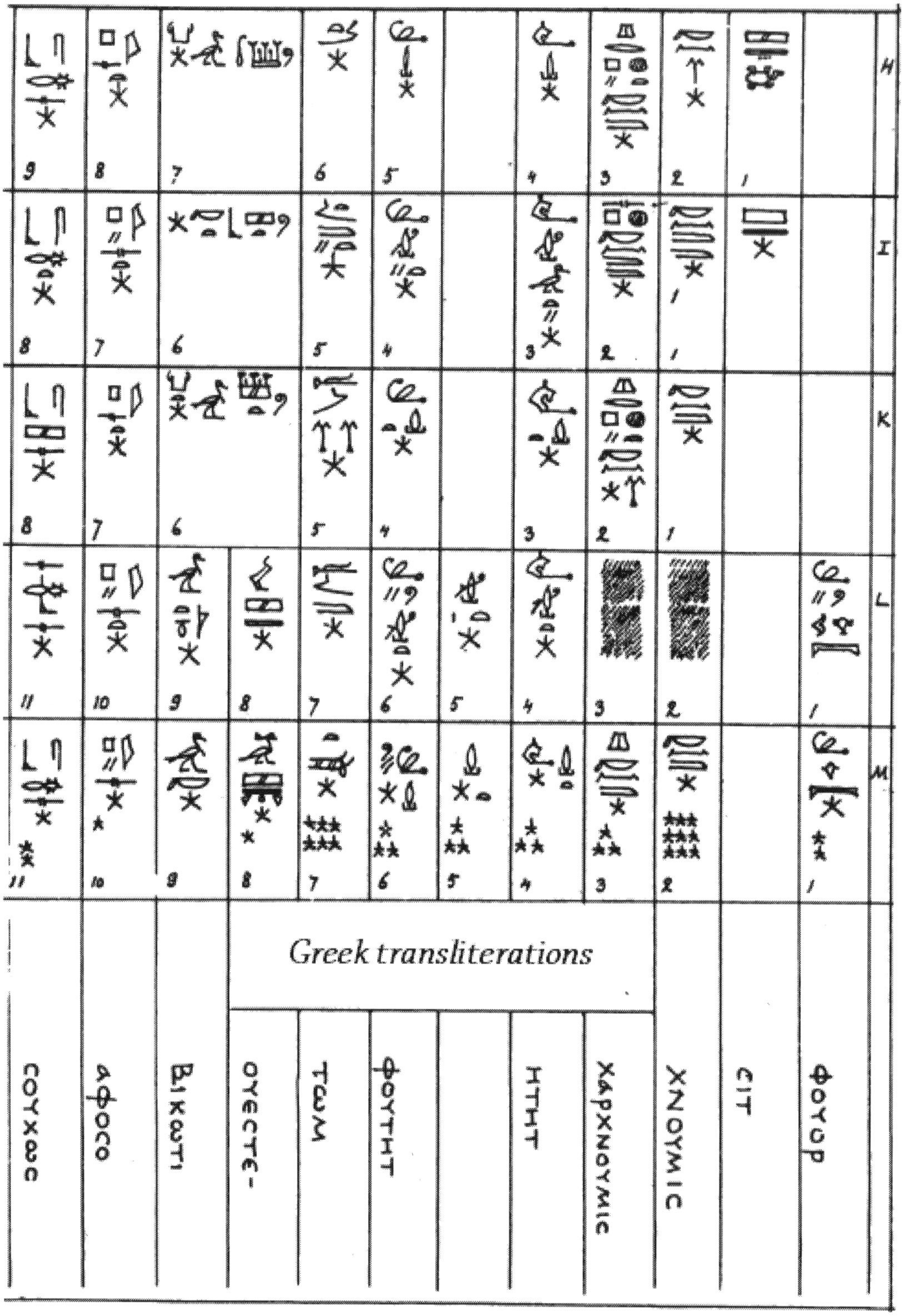

The lists of decans of the Greco-Roman period [continued]

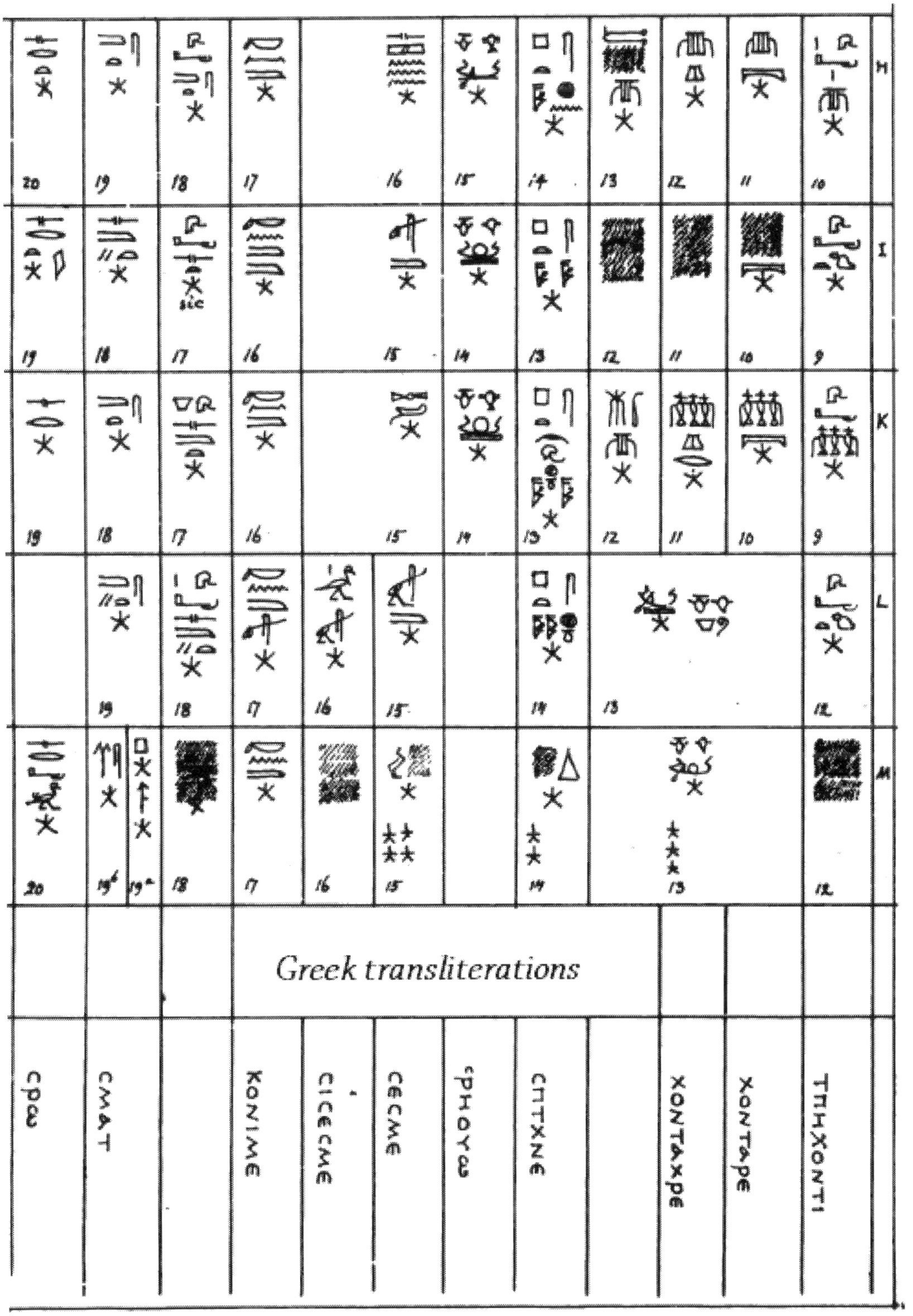

The lists of decans of the Greco-Roman period [continued]

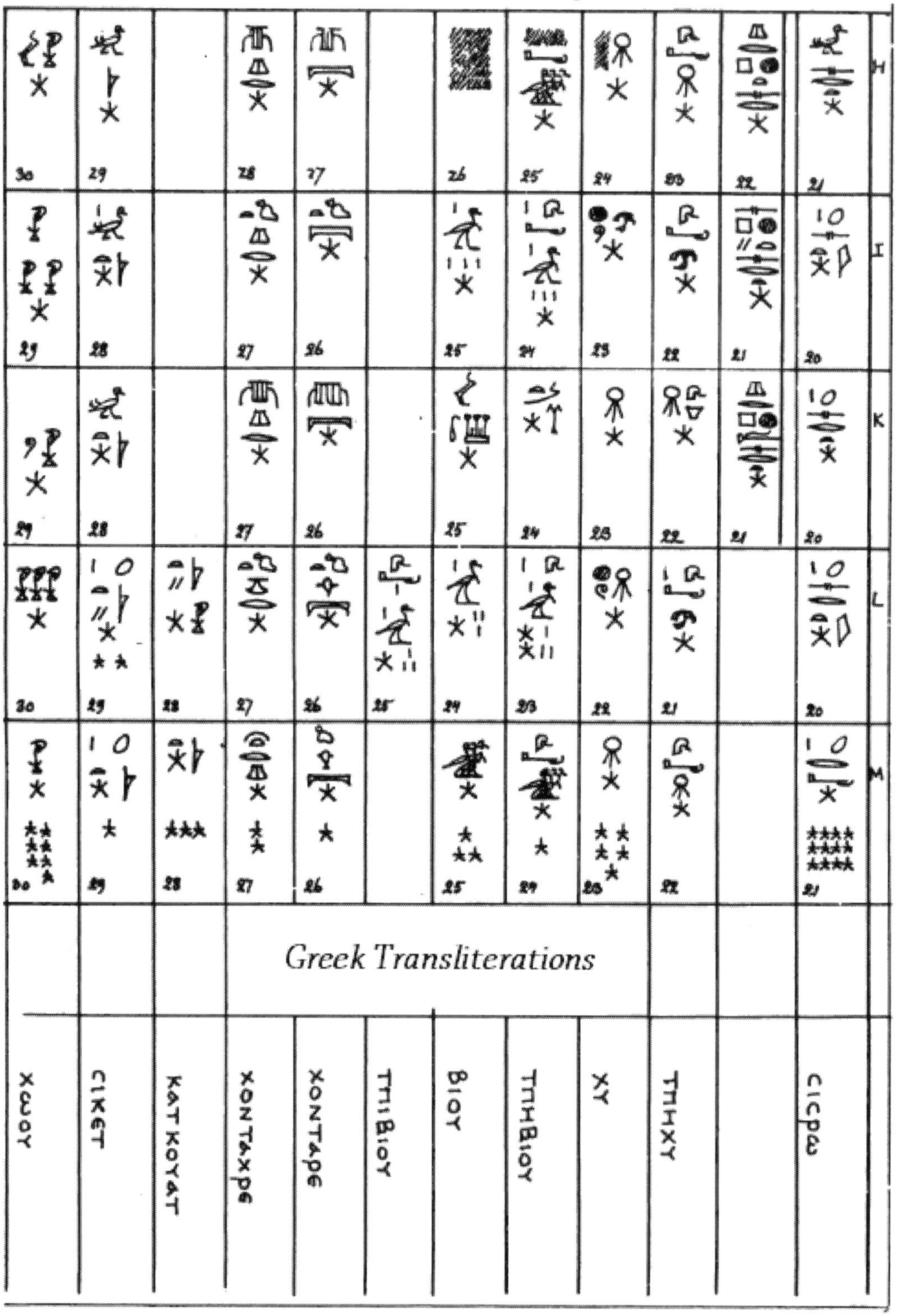

The lists of decans of the Greco-Roman period (end)

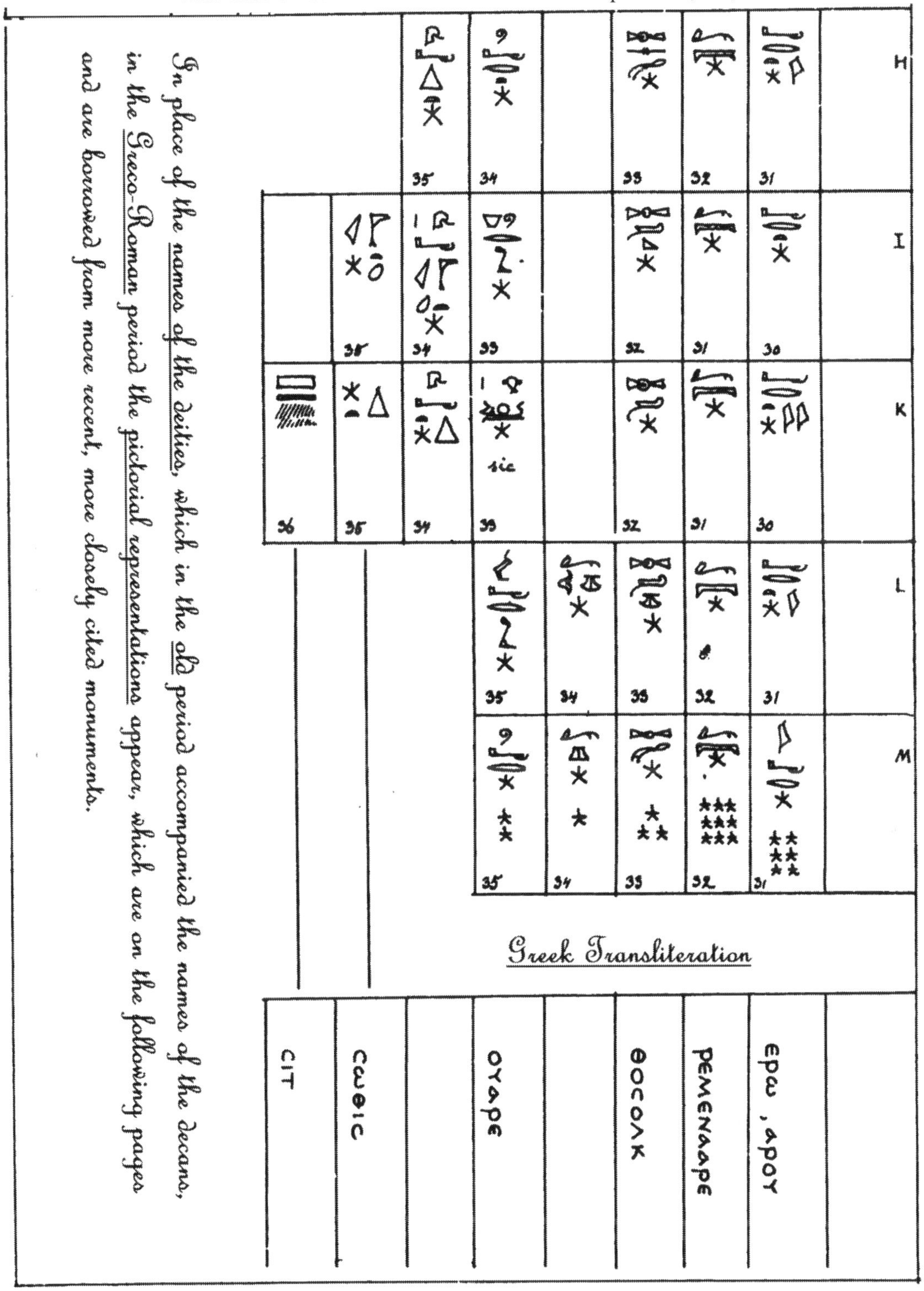

The Decan Images from Roman Times

[compare L and M]

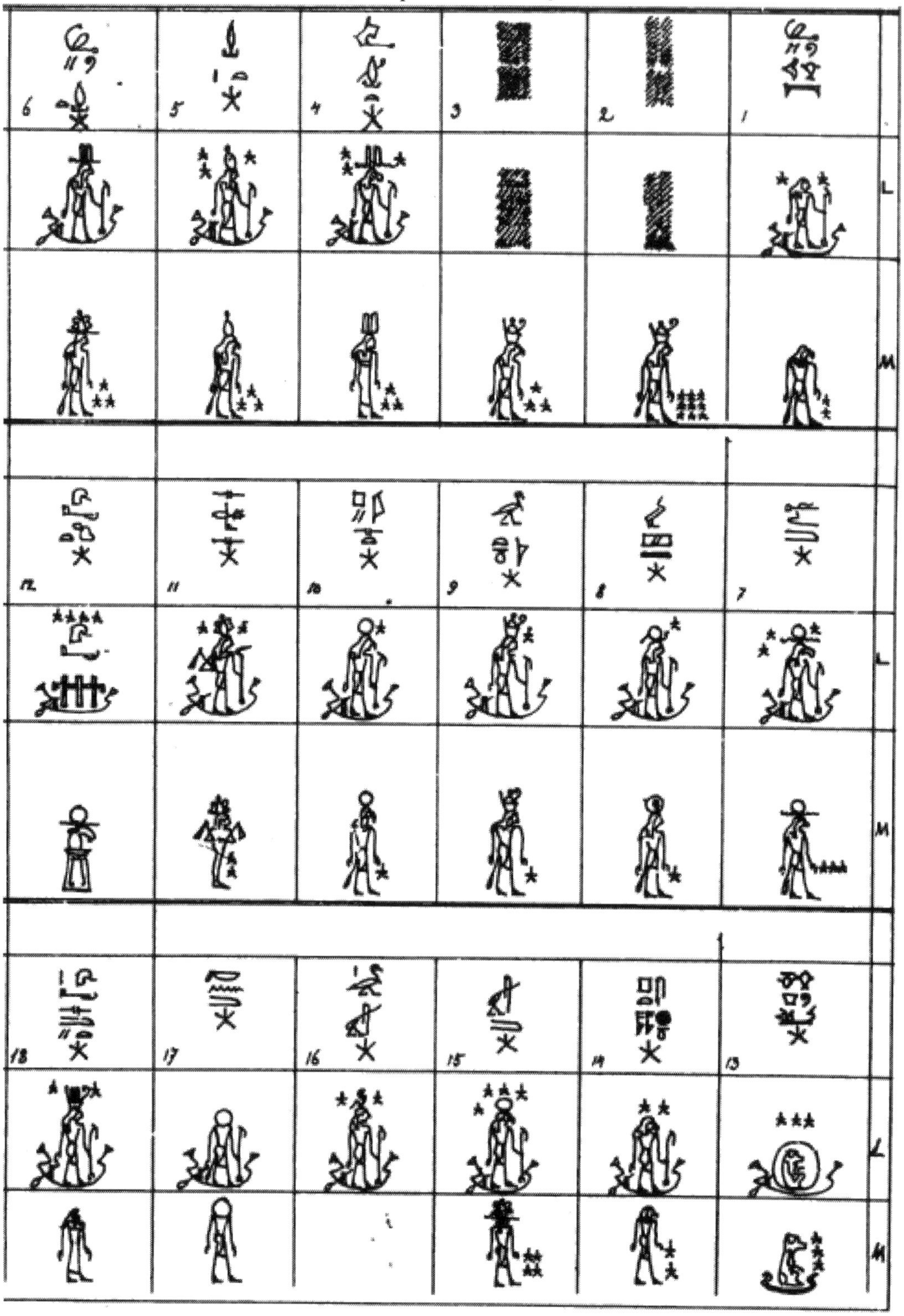

The decan pictures from Roman times [end]

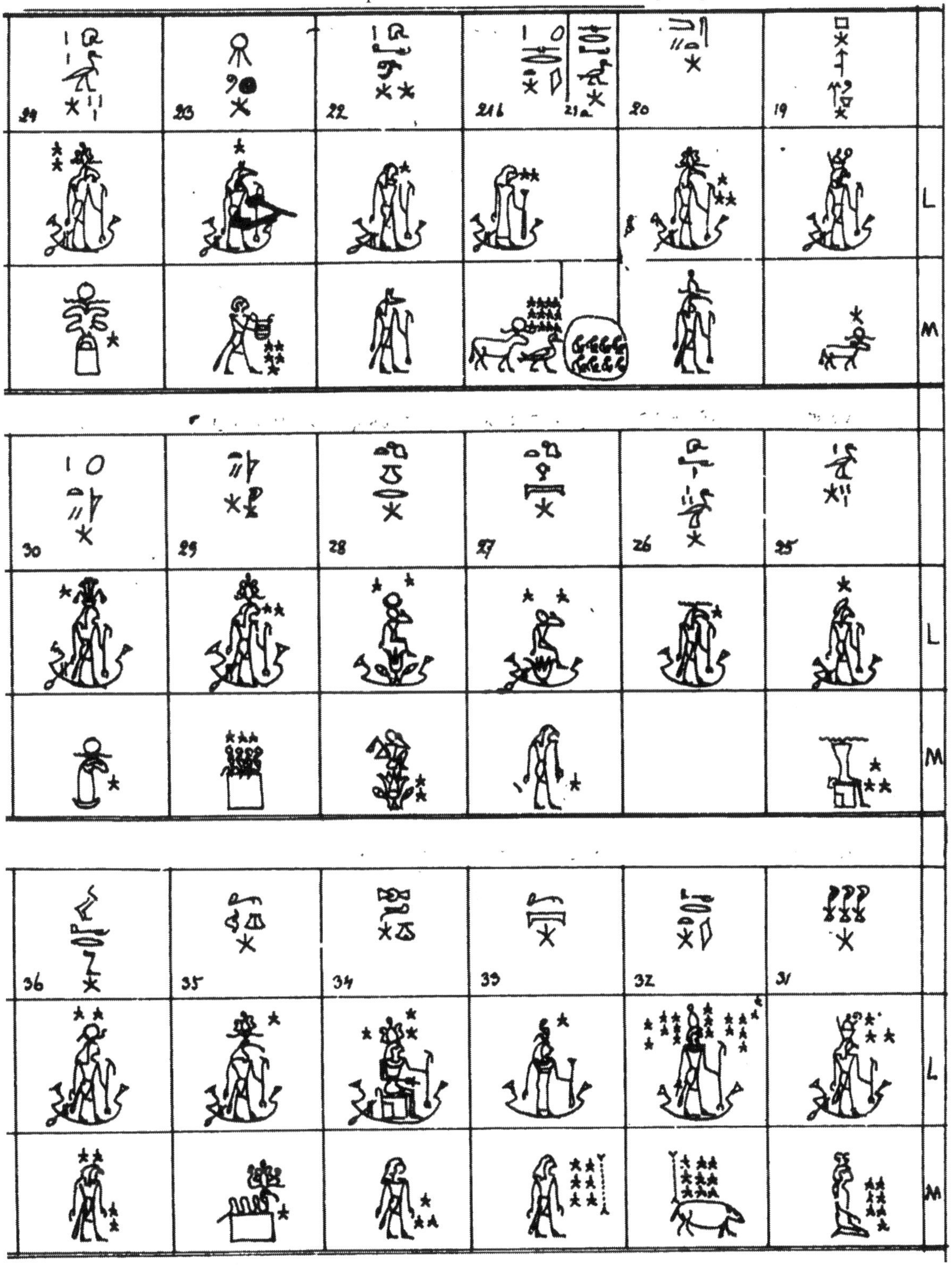

End of the decan lists.

Text on the above decan comparisons

I. *The decan lists of the older period*

The listed decans belong to the following constellations:

1. *kenmut*
2. *tat*
3. *θmet* or
4. , , *usθa*, *usθa*, *usti* or *besθet.*
5. , , , , *bokθa*, *bokθa*, *bekati*, *bekaθ*

[6. *sibu mahu*, inserted in the lists from the 20th dynasty]

6. , , *Xontet*, *Xonti*, *Xont.*
7. *sapet Xonnu*
8. *her-ab-ua*
9. , written purely phonetically: *sesmu* (and the varr. *sasmu*, *sesmu*)
10. *kenemu*, *kenem*
11. *smad*, *smadet* (i.e. halver, with the sign of the crescent behind it, which will return in the group for the 15th day of the lunar month, compare above, page 40, 15 under e', e'''), on it.
12. *seret*, , *sat*, *sat* "the sheep"
13. *Xu* order *XuXu* (actually: the shining one or the two shining ones)
14. *bibi*, *bibiu* (i.e. "the two souls")
15. , , *Xont*, *Xont* (see above no. 6)
16. *kod*
17. , , *aret*
18. , *Xau*, *Xa* "thousand star"
19. , , *sah*, the constellation of Orion
20. , *Sopdet*, the constellation of Sothis, or Sirius.

A noticeable difference in the decan lists of the 19th and 20th dynasties is evidenced by the presence of the new decan of the 'full stars' (, see above under [6]), which appears only in the lists of the epoch of the 20th dynasty.

That the names of the decans referred to real pictures in the Egyptian starry sky is proven by the actual existing and verifiable figures as well as the designations which, depending on the location, referred to the individual parts of larger constellations and which are contained in the following list.

, <u>*tpa*</u> "<u>*head, top, uppermost part*</u>," as opposed to , , <u>*Xer-Xeped*</u> mentioned body part, probably meaning the naval region of the abdomen.

, <u>*hr*</u>, <u>*hru*</u> "<u>*upper*</u>" as opposed to , <u>*hr-ab*</u>, "<u>*medium*</u>, <u>*middle*</u>" and , <u>*Xr*</u>, <u>*Xru*</u> "<u>*lower*</u>".

, <u>*hat*</u>, "<u>*front*</u>, <u>*forepart*</u>" as opposed to , <u>*phui*</u>, "<u>*rear, hindquarters*</u>".

What <u>*sisi*</u> [in the Greek decan list C1 -] means is unclear to me.

The , which occurs only once, means something like a tablet, a panel, and probably refers to the wooden panel serving as a base for a standing or seated figure. The relations to certain parts of the body appear most clearly in the decan designations, which refer to the constellation of Orion; so in, , <u>*rmn hr*</u>, <u>*hr rmn*</u> "the upper arm" (Greek, compare Lepsius, *Die altägypt. Elle*,[51] p. 34), , , , <u>*rmn Xr*</u>, <u>*Xr-rmn*</u> "the forearm," "the hand," <u>*master*</u> "the ear," <u>*uaret*</u> "the foot."

---

[51] <u>Translator note:</u> Lepsius, *Die altägyptische Elle und ihre Einteilung* (or, *The ancient Egyptian Cubit and its Classification*)

A. 1Tpā Kenmut	2. Kenmut	3. Xr-Xpd-	4 ḫā-tat
B 1Tpā Kemut	2 Xr-Xpd-Kemut	3 Kenmut	4 ḫō-tat
C 1Tpā Kenmut	2. Kenmut	3 Xr-Xpd-Kenmut	4 ḫātat
D 1..ā Kenmut	2. Kenmut	3 Xr-Xpd-Kenmut	4 ... tat
E	2	3	4 ... tat
F. 1Tpā mui	2. Kenmut	3 Xpd-n-Kenmut	4 .. 1-taīt
G 1Tpā Kenmut	2 Xr-Xpd-Kenmut	3 Kenmut	4 ḫā-tat
"Top of Kenmut"	"Kenmut"	"Naval region of Kenmut"	"Front of Tat"

A 5 pḫu-tat	6 θomet-ḫrt	7 θomet-Xrt	8 ušθà
B 5 pḫu-tat	6 θomet-ḫrt	7 (θomet)-Xrt	8 ušθ
C 5 pḫui-taīt	6 tomet-ḫrt	(7) Xrt	7 ušti
D 5 .... ta	6 tomet-ḫrt	7 tomet-Xrt	8 ušti
E 5 pḫui-tat	6 tomet-ḫrt	7 tomet-Xrt	8 ušti-bokti
F 5 .... taī	6 tnnes-ḫr	7 tnnes-Xr	
G 5 pḫui-tat	6 θomet-ḫrt	7 tomet-Xrt	8 bešθet-bekaθ
"Backside of tat"	"Upper box"	"Lower box"	

A 9 bokθà		10 tpā-Xontet	11 Xontet-ḫrt
B 9 bokθ		10 idem	11 Xont-ḫrt
C 8 bekati	9 sibu maḫu	10 idem	11 Xontet-ḫrt
D	9 sibu maḫu	10	11 Xonti-ḫrt
E	9 sibu maḫu	10 tpā-Xont	11 Xont-ḫrt
F bekθ pḫu		tpā-Xont..	Xontet-ḫrt
G	9 sibu-maḫu	10 tpā-Xontet	11 Xontet-ḫr
	"Full stars"	"Top of Xontet"	"Upper Xontet"

A	12 Xontet-Xrt	13 Bemas n Xontet	14 Sapt Xonnu	15 ḥr-āb-ua
B	12 idem	13 idem	14 sa(p) n sutn Xon	15 idem
C	12 idem	13 ṯi-Xont	14 sapti Xonnu	15 idem
D	12 Xonti-Xr(t)	13 Bemas n Xont(i)	14 ..p..... nu	15 idem
E	12 Xont-Xrt	13 Bemas n Xont	14 sapt Xonti	15 idem
F	„Stern der Sed"		sapti Xonnu	ḥr-āb-uaf
G	12 Xontet-Xrt "Lower Xontet"	13 ...s n Xont "Panel of Xontet"	14 .... Xon "Edge of Xonnu"	15 ḥr-āb-ua "middle of the ship"

A	16 Šesmu	17 Kenmu	18 Smadet	19 ṯpā-smad
B	16 Šesmu	17 Kenem	18	19 smad
C	16 Sasmu	17 Kereman..	18 ṯpā-smad	19 ibid.
D	16 Šesmu	17 Kenmu	18 ṯpā-smaB	19 ibid.
E	16 Šesmu	17 ibid.	18 ṯpā-smad	19 ibid.
F				
G	16 Šesm(u) "The braid"	17 Kenem "Kenemu"	"The crescent moon"	18 smad "Tip of the crescent"

A	20 sart	21 sisi-sart	22 Xr-Xpd-sart	
B	20 sàt	21 sisi-sàt	22 ibid.	23 ṯpā-Xu
C	20 ✱ ∘∘∘ ∘∘∘	21 sisi-sart	22 Xr-Xpd-sar	23 Xu 2
D	20 ..r.	21 sisi-....t	22 idem	23 ṯpā-XuXu
E	20	21	22 Xr-Xpd-sar..	23 ṯpā-XuXu
F				ṯpā-Xu
G	19 sat "The sheep"	20 sart sisi "The 2 ..... sheep"	21 Xrt-sart "Naval of the sheep"	22 ṯpā-XuXu "Top of the 2 Xu"

A	23 XuXu	24 bibi	25 Xont-ḫru	26 ḫr-àb-Xon[t]u
B	24 XuXu	25 bibiu	26 ḫru-Xont	
C	24 Xu 2	25 bibi	26 Xont-ḫr	27 ḫr-àb-Xont
D	24 XuXu	25 bibi	26 Xonti-ḫru	27 Xonti-ḫr-àb
E	24 XuXu	25 bibi	26 Xont-ḫru	27 ḫr-àb-Xont
F		biu		
G	23 XuXu "The two Xu"	{24 toā-bibi {25 bibi "The two souls"	26 Xont-ḫrt "Upper Xont"	27 ḫr-àb-Xontet „Middle of Xont"

A	27 Xont-Xru	28 Kod	29 sisi Kod	30 āret
B	27 Xru	28 Kod	29 ibid.	30 āret
C	28 Xont-Xr	29 	30 šesui bibi	31 Xa
D	28 Xonti-Xru	29	30 sisi-Kod	31 Xau
E	28 Xont-Xru	29 Kod	30 ibid.	31 Xau
F				
G	28 Xont-Xrt "Lower Xont"	29 Kod „Kod"	30 ibid. "the 2 . . . of the Kod"	31 Xau „āret"—„1000 stars"

A	31 Xau	32 rmn ḫru àn saḫ	33 masler saḫ	
B		31 Xr-rmn-saḫ	32 	33 rmn saḫ
C	32 āret	33 àler saḫ, 7 Sterne	34 uāret Xr	35 tot-(ā?) Xr
D	32	33	34	35 ....
E	32 āret	33 àn (àler?) saḫ		34 ....ḫr....
F		Xr-rmn n sa		1pā-saī
G	32 ār.. „1000 stars"—āret	33	34	

A	34 rmn-Xr sah	35 ā-sah	36 sah	37 Sopdet		
B	34 hr-rmn-sah			35 Isis-Sopdet		
C	36 rmn-Xr-sah	37 ā-sah		38 Sopdet	39 6a	40 "sāt of the Sah under the foot of Sah."
D	36 rmn	37 ā-sah	38 sah	39		
E	35	36	37 sah, the star of Osiris under ā-sah	38 Sopdet, the star of Isis		
F			sai			
G	35 rmn-Xr....	36 ā-[sah]		37 Sopdet-Isis		

The Dieties of the Decan Realms in the Astronomical Sense (see p. 111)

	Dekan	A'	B'	C'	D'
1.	Ipā-kenmut	Seb	Hapi Asmet	....	Hapi Amset
2	Kenmut	Bi	Isis	....	Isis
3	Xr-Xpd-kenmut	Xontet Xast		....	The Horus children
4.	hā-tat	Isis	Duamutef	The Horus children	ibid.
5	phui-tat	"The Lady of Aphroditopolis"	The Horus children	ibid.	ibid.
6	tomet hrt	àmsetà Hapi	Duamutef	ibid.	ibid.
7	tomet-Xrt	Qebhsonuf		Duamutef	Duamutef
8	usθà	Duamutef	Duamutef	Duamutef	ibid.
9	bekati	Duamutef	Hapi		
		Qebhsonuf			
	[sibu mahu]			Duamutef	Duamutef
10	Ipā-Xontet	Duamutef Hapi	3 Horus	Duamutef Hapi	Duamutef

	Decan	A'	B'	C'	D'
11	Xontet-ḥrt	Horus		Horus	.....
12	Xontet-Xrt	Set			
13	Θemat n Xontet	Horus	"missing"	"missing"	"missing"
14	Sopt-Xonnu	Isis & Nephthys	Isis & Nephthys	Isis, Lady of Busiris	"missing"
15	ḥer-àb-ua	Set	urà	"missing"	"missing"
16	Šesmu	Horus		"missing"	"missing"
17	Kenmu	àmseθ, Ḥapī Duamutef Qebḥsonuf	the Horus children	the Horus children	the Horus children
18	Ìpā-smad	Horus	.....	Horus	Horus
19	smad	Ḥapī	Ḥapī	Ḥapī	Ḥapī
20	saret	Isis	Isis	Isis	
21	sisi-saret	Duamutef Qebḥsonuf	Duamutef	Duamutef	
22	Xr-Xpd-saret	Qebḥsonuf	Qebḥsonuf	Qebḥsonuf	
23	ìpā-Xuxu		Duamutef	Duamutef	
24	Xuxu	Duamutef Qebḥsonuf	Duamutef Qebḥsonuf	Duamutef	
25	bibi	amseθà Ḥapī	Ḥapī àsmat	amseθà	
26	Xont-ḥru	Horus	the Horus	Horus	
27	ḥeràbXont	Horus		Horus	
28	Xont-Xru	Horus		Horus	
29	Kod		Ḥapī		
30	sisi-Kod	Amseθ Ḥapī Qebḥsonuf Duamutef Qebḥsonuf Ḥapī		Ḥapī Qebḥsonuf	

	Dekan	Á	B́	Ć	D́
31	Xau	Hapī			
32	āret	amseθà	Mat-ḫur	Ḫur (Horus)	
33	rmn ḫru àn Saḫ	Duamutef	mas-ḫur		
		Qebḫsonuf			
34	máster Saḫ	Mat-ḫur Ḫur			
35	rmn xr Saḫ	ibid.			
36	ā- Saḫ	mat-ḫur			Mat-ḫur
37	Saḫ	mat-ḫur Osiris			

Note on the meaning of the group *Xet-mu*

The decans in the above list show individual examples in which the group *Xet-mu* can be found next to and in addition to the names of the gods listed. As Prof. Dr. Dumichen correctly recognized, the word *Xet*, originally *the belly*, meaning *the body*, in the so-called Kyphi recipes as a *mass* composed of several ingredients. As the result of a composition of liquids it is *Xet-mu* "a liquid mass", of dry ingredients it is or *Xet suu* a "*dry mass*" (see *Ztsch.*[52] 1879, p. 97, ff. especially lines 13, 15, 19, 10, 22 and Mar. *Dend.*[53] I, 47, e). As I said in the *Wort.*[54] Suppl. p. 965, (plur.) also serves to denote a conglomeration of stars, a star cluster, as we are accustomed to say, when the individual stars can still be clearly distinguished, as opposed to the indissoluble nebulae. Such star clusters appear in our decan lists as follows:

1. Decan 20: *sarb* "the sheep." In addition to the Isis name (var.), there is a legend in B', improved in *Xet mu nt* i.e. "*the star cluster of*" [i.e. Sheep, which stands next to it in the picture and clearly carries 5+4=9 star balls, arranged in 2 rows].

[52] Translator note: *Zeitschrift für ägyptische Sprache und Altertumskunde*(?)
[53] Translator note: Mariette, *Dendérah : description générale du grand temple de cette ville.*
[54] Translator note: *Worterbuch*; i.e., dictionary

2. Decan 27:  *her-ab-Xont.* In addition to the Horus name  or 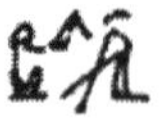 in D' the legend: 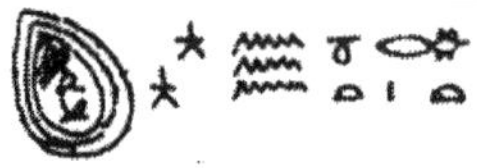 "*the star cluster of...*" (accompanied by a pear-shaped figure that returns in C' as ).

3. Decan 29: 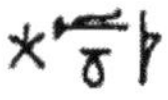 *kod.* Instead of the name of the deity in question, in A' next to the figure  there is the legend: 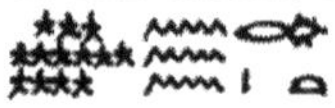 "*the star cluster of* [decan Kod]", in C' the legend "*the star cluster of the 13 stars,*" in D' on the other hand, *close to the front part of the Orion* ship, the following inscription: "the star cluster of the eight stars." After these remarks, it will not be difficult to fully appreciate the passage from the great Harris Papyrus No I, which I have given in the *Wörterbuche* Suppl. p. 965, and which I will repeat below: 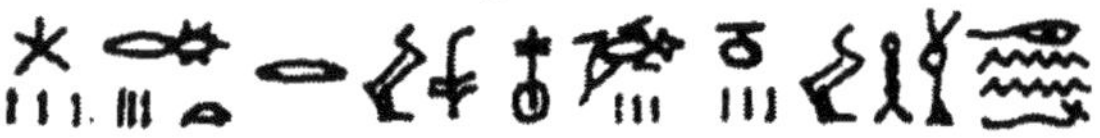 "I have had vessels made for him *in greater numbers and more glorious than the star clusters.*"

H		1. šela	2 kenem	3 χr-χpti-Kenem
I		(ḏ) šela	1 Kenemm	2 δχp-Kenemm
K			1 Kenem	2 χr-χpti-Kenem
L	1. Phui-ḫr		2	3 ....
M	1 Phui-ḫr		2 Kenem	3 χr-Kenem
H	4 ḫā-lá	.....	5 phui-lá	6 tom
I	3 ḫā-uláti	......	4 phui-uláti	5 tommti
K	3 ḫā-lát	....	4 phui-lát	5 ṭommem
L	4. ḫā-ulát	5 ulát	6 phui-ulát	7 ṭomm
M	4 tál-ḫā	5 lát	6 phui-lá	7 tom
H	7. ušθ-bika	8 áposet	9 sebχes	10 lpā-χont
I	6 ušl-bekat	7 ápiset	8 sebχes	9 lpā-χont
K	6 ušt-bikat	7 áposet	8 seb šes	9 lpā-χont
L	8 ušla 9 bikot	10 ápiset	11 sebχes	12 lpā-χont
M	8 ušla 9 bik	10 ápis	11 sebχes	12 ....
H	11 χont-ḫr	12 χont-χr	13 θ.... χont	14 spt-χon
I	10 .....-ḫr	11	12	13 spt-χon
K	10 χont-ḫr	11 χont-χr	12 θemas-χont	13 spt-χon
L	13 ḫr-áb-uā			14 Spt-χon
M	13 ḫr-áb-uā			14 sopt-.....

H	15 ḥr-áb-ua	16 sešmu		17 Kenem
I	14 ibid.	15 sem. (sešm?)		16 Kenemm
K	14 ibid.	15 θosālq		16 Kenem
L		15 sem (sešem?)	16 si-sem (si-sšm?)	17 Kenem
M		15 ....u	16	17 Kenem
H	18 ɩpā-smat		19 smat	20 saret
I	17 ɩpā-s[m]at		18 smati	19 sarát
K	17 ɩpā-smat		18 smat	19 sar
L	18 ɩpā-smati		19 smati	
M	18	19a „der Abendstern"	19b sma	20 sarā
H	21 si-saret	22 Xr-Xpt-sar	23 ɩpā-Xu	24 Xu
I	20 si-sát	21 sXpti-saret	22 ibid.	23 ibid.
K	20 si-saret	21 Xr-Xpt-saret	22 ibid.	23 ibid.
L	20 si-sarát		21 ibid.	22 ibid.
m	21 si-rā (sic)		22 ibid.	23 ibid.
H	25 [ɩp]ā-biu	26	2°	27 Xont-ḥr
I	24 ɩpā-biu	25 biu		26 ibid.
K	24 ɩomm	25 ušeθ		26 ibid.
L	23 ɩep-biu	24 biu	25 ɩpā-biu	26 ibid.
M	24 ɩpā-biu	25 biu		26 ibid.
H	28 Xont-Xr		29 si-Kod	30 Xau
I	27 ibid.		28 si-Kot	29 ibid.

K	27 Xont-Xr		28 si-kot	29 Xau
L	27 ibid.	28 koti-Xa	29 si-koti	30 ibid.
M	27 ibid.	28 kot	29 si-kot	30 Xa
H	30 ārāt	32 rmn-hr	33 θos-ālq	
I	30 ārel	31 ibid.	32 ibid.	
K	30 āril	31 ibid.	32 ibid.	
L	31 ārāt	32 ibid.	33 ibid.	34 rmn-Xr
M	31 āār	32 ibid.	33 ibid.	34 rmn-Xr
H	34 uāret	35 tpā-sopdet		
I	33 uār	34 ibid.	35 sopdet nutār	
K	33 hr-ua	34 ibid.	35 sopdet	36 Šesa
L	35 uār			
M	35 uār			

Comments on the above list. The difference between the rows of decans of the older and more recent periods is that the following new decan designations appear in the Greco-Roman epoch: phui-hr, seta (see above, p. 86), aposot and sebXes (see above, see p. 87), tpa-biu, θos-alq, uar and tpa-sopdet. It is also based on the merging of two older decan names into a single newer one, such as in tom instead of tomet hrt and tomet Xrt, in ust-bkt in place of and usti and bkati, in the replacement of older decan names or rather their representation by more recent designations, such as in the series tpa-kenmut, sibu mahu, hr-ab-Xont, and in those with the Egyptian naming of Orion, sah, sahu, compound names. Other differences, especially in the changed order, are easy to recognize for yourself. The series of Egyptian decan names preserved from Salmasius joins the lists of the more recent epoch. To complete the same, and for the sake of better comparison, I quote the following 4 decan names from the temple of Ombos (Ptolemy epoch), which appear in connection with daily hours (see Champollion, *Not. Desc.*[55] I, p. 237) and they appeared in the following legends:

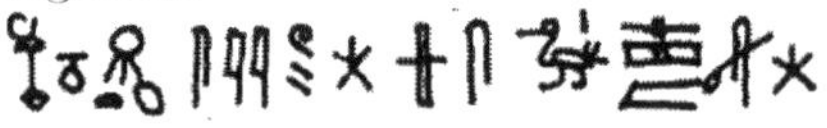 <u>*ubenet siui amos si-sesem*</u>, the first "hour of the day". The star in it is <u>*Si-sesem*</u>."

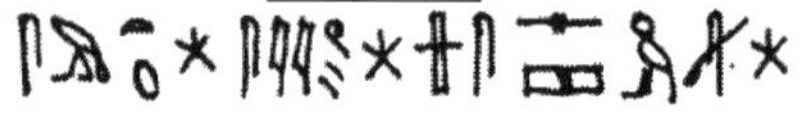 <u>*mak-n-nebes siui amos Kenein*</u>."

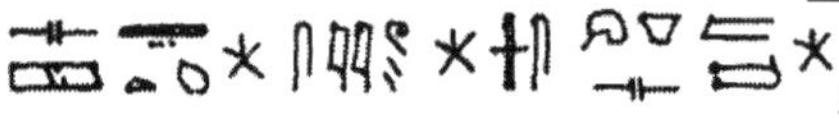 <u>*sestat siui amos tpa-sma*θ</u> "the fourth hour of the day. The star in it is <u>*Tpa-sma*θ</u>."

The list obtained from Salmasius, which Lepsius first drew attention to the importance of (see introduction to *Chronologie der Ägypter*, p. 70), according to Champollion, has its complete and absolute confirmation by the following combination of the Greek and Egyptian names, with special consideration to that of Goodwin in the "*Melanges,*" 1864, p. 294 ff. based on an Egyptian horoscope written in Greek.

[55] <u>Translator note</u>: Champollion, *Monuments de l'Égypte et de la Nubie. Notices descriptives.*

[VI Sagittarius 18. Κονιμέ Egypt. konime

VII Capricorn 19. Σμάτ Egypt. smat

20. Σρώ q. Cρωι Egypt. srȧt

21 Σισρώ q. Cιcρωι Egypt. si-srȧt

VIII. Aquarius 22. Τπηχύ Egypt. ipā-χu

23. χύ Egypt. χu

24. Τπηβίου Egypt. ipā-biu

IX Pisces 25. Βίου Egypt. , biu

26. Χονταρέ Egypt. χont-ḥar

27. Τπιβίου Egypt. ipi-biu

X Aries 28. Χονταρέ Egypt. see № 26.

29. Χονταχρέ Egypt. χont-χre

30. Σικέτ Egypt. si-ket

XI Taurus 31 Χώου Egypt. χau

32. Ἐρῶ q. Αρου Egypt. ārȧt

33. q. Ρεμενααρέ Egypt. remen-ḥare

XII Gemini 34. Θοσόλκ Egypt. θos ālq

35. Οὔαρε Egypt. uāret

36. φουός Egypt. phu-ḥor.

The Rise of the Decans

The rising of the decan stars, noted at ten-day intervals (the so-called decades), which according to the nature of the old Egyptian transitional year, consisting of 365 days, had to fall one year on the 1st, 11th and 21st day, in the following year on the 6th, 16th and 26th day of each month, are found in more detail and in context only on two monuments, which Lepsius had first recognized correctly according to their astronomical meaning (see "Introduction to the *Chronol.*"[56] p. 115 ff.) One concerns the rise of the Decan stars, which are listed on the ceiling of a hall in the tomb of King Ramses IV at Thebes, on and below the body of the sky goddess, albeit with some (easy to check) errors in detail, which are the burden of the copist. The table below contains the transcript of all legends, as I have recorded them on the spot, with reference to the depiction of the heavenly figure published by *Lepsius* in the "*Wall Paintings*"[57] of the Egyptian Museum in Berlin on plate 7, its bearer (the god *Su*, Σῶς) and other related figures. My deviations in individual cases mainly concern the numbers in the heiroglyphic inscriptions. They also contain corrections of individual errors in the characters, which may have mistakenly crept into the excellent publication and have been overlooked. The letters and numbers on my copy (see the table) will help you understand the following description and translation of the legends.

The rising epochs of the decans are, as Lepsius first demonstrated, characterized by  *tepet*, *early rising*,  *sa-duat*, *midnight rising*, and  *maset*, *late rising*, by day and month. The interval (Egyptian  *amtu en son*, i.e. "what lies between 2, the gap of 2") between the early rising and the midnight rising of each decan is 90 days according to the theoretical layout of the table, the interval between the midnight and the late rise is 70 days, or in total between the early rise and the late rise 90+70=160 days. As Dr. *Fr. Gensler* (*Ztschft.* 1872,[58] p. 61) has shown from the astronomical point of view, in the transcription of the annual observations of stars, the first large visibility interval for the horizon of Thebes is actually 160.4 days. It goes without saying that the starting epochs of the decans can be calculated theoretically according to this, where, first and foremost, the data relating to the decans [*Kenemut* (No. 2), *Xr-Xpd-kenmut* (No. 3), *hat-tat* (No. 4), *phu-tat* (No. 5) and *tomet hrt Xrt* (No. 6)] next to the feet of the celestial bearer (see A-B-C) give the important starting point of the calculations. The inscription at F - according to the calculation, necessarily refers to the decan [*usti*, no. 7], although the same is not listed by name. The text at I leads directly to the rising of the decan  *be-kati* (var. instead of 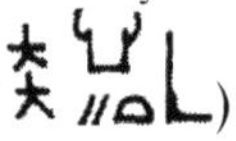) or No. 8 of the decan series, whereby it should be noted that the inscriptions under H and G repeat the related data again. The text at G also deserves special attention, because it expresses the Egyptian terms ,  and  for the three risings, the heliacal, midnight and late risings, in a way that seems to be explanatory, and just as an example the risings epochs of the decane *bekati* are chosen:

---

[56] Translator note: Lepsius, *Die Chronologie der Aegypter.*

[57] Translator note: Lepsius, *Koenigliche Museen, Abtheilung der Aegyptischen Alterthümer die Wandgemälde der verschiedenen Raeume*

[58] Translator note : Most likely the journal *Zeitschrift für ägyptische sprache und altertumskunde*, Volume 10.

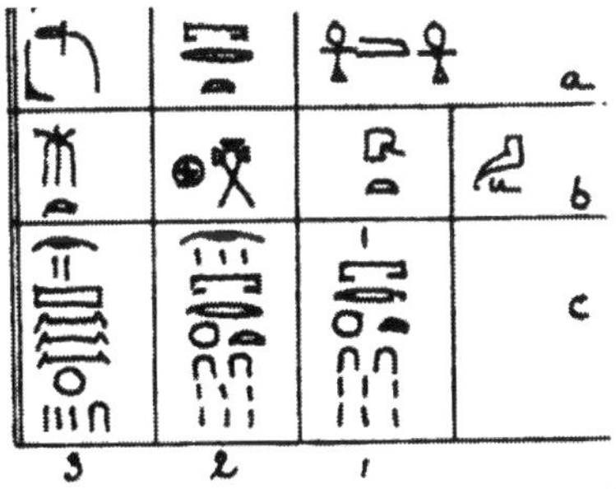

i.e. (a) "the *coming into life* (*anX*) as *decan* (*anX*, see previous p. 102), the *appearance* (*piret*) and the *dissolution* (*uab*)
"(b) [entering] as *tepel* (early morning), midnight rising (*sa-duat*) and late rising (*maset*)
"(c) 26/5, 27/7 and 10/10" [incorrect on Stella as 26/5, 26/8 and 6/11]. In the description of the dates I have used the formula x/y, where x expresses the day of the month, y the arrangement of the month in the course of the year.

The texts under D and E contain the rising epochs of the following 36-7=29 decans up to the end, with decan *tpa-kenmut* (No 1 of the index L-M) with the early rising on 16/3 the 36th position and the Sothis decan  *sopdet* with the early rising on 6/3 takes the 35th place. Since in the theoretical version of the table the 5 intercalary days at the relevant epochs of ascent are not taken into account, it necessarily follows that the decan  *a-sah* referring to them must also be discarded. I also remark, with regard to the last decans (no. 33-38), to which the dawns from 6/2 to 6/3 refer, that in the works of Champollion (*Mon.*[59] III, 275) and Rosellini (*Mon. d. Cult.*[60] 68), and copies published after them by Lepsius (*Wandgemälde*, Plate 7) show deviations from my own copy, as can be seen from the following compilation of the copies:

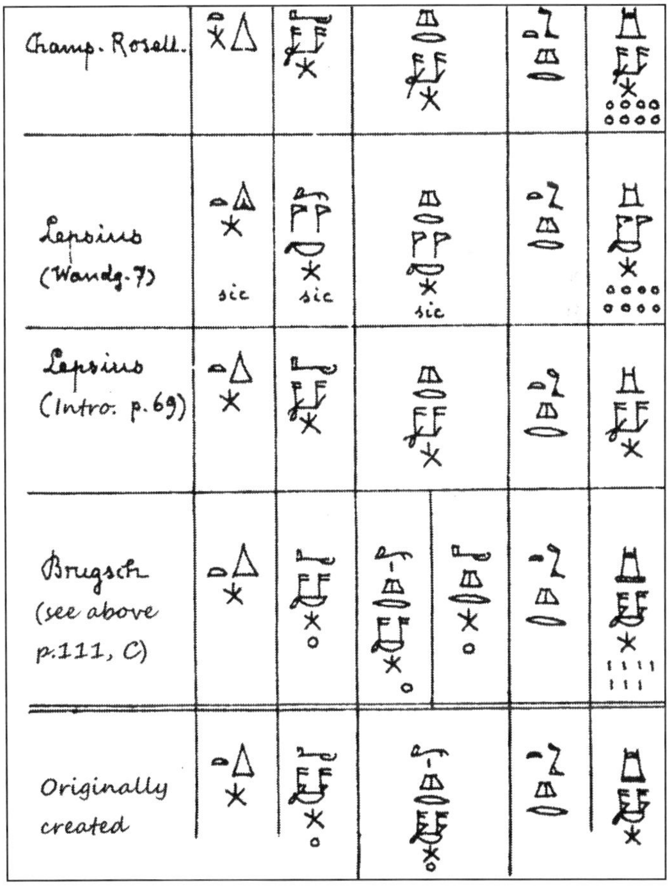

A comparison of these copies with each other leads to the result that I have summarized below as: "originally created." If one compares the corresponding decans from the Greco-Roman period with this corrected small list, the following instructive compilation results.

[59] Translator note: Champollion, *Monuments de l'Egypte et de la Nubie.*
[60] Translator note: Rosellini, *I monumenti dell'Egitto e della Nubia. Pt. 3. Monumenti del culto.*

Older List C	Greco-Roman List	Salmasius
àter sah	rmn hr	33. Ῥεμενάρέ
uāret Xr	θos-ālq	34 Θοσόλκ
rmn Xr	uāret	35 Οὔαρε
ā-sah	tpā-sopdet-od. phu-hr	36 Φουόρ
sopdet	sopdet	1 Σωθίς
tpā-Kenmut	šit	2 Σίτ
Kenmut	Knm	3 Κνευμίς

One sees from this that the old decan *tpa-kenmut* (No. 1 of the ascending) was represented in the newer epoch by the old constellation *sit* (see above p. 86), also written šeθu, šiθu (see p. 85, B), "the turtle, the two turtles".

The connection between the decan *tpa-kenmut* and the constellation šit or šiθu is actually also proven by the representations of the rising decan from the epoch of the 20th dynasty. Because exactly at the point where one should expect the indication of the rising epochs of the 1st decan of the *Tpa-kenmut* list, there are two sun-like circles (i.e. 2 stars of the first magnitude) in close proximity to the feet of the celestial figure (i.e. at the east point of the representation) and above it the inscription i.e. *anX siθu* "*the early rise of the (two) turtles.*"

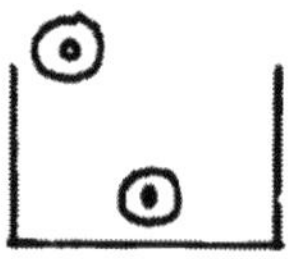

Table of the decan's rise from the times of the 20th Dynasty

	Decan name	Early rise (Inscript. ¦ theoret.)	Middle rise (Inscript. ¦ theoret.)	Late rise (Inscript. ¦ theoret.)	Place on the board
1	2. Kenmut	26 4 ¦ 3	26 6	6 9	A, 1–3
2	3. Xr-Xpd-Kenmut	6 4	6 7	16 9	B, 1–3
3	4. ḥāt-tāt	16 4	16 8 ¦ 7	26 9	C, 1–3
4	5. pḥu-tat	26 4	16 ¦ 26 7	16 · 6 10	C, 1–3
5	6. tom ḥrt Xrt	6 5	6 7 ¦ 8	26 ¦ 16 10	C, 1–3
6	7. ušti	6 ¦ 16 6 ¦ 5	6 ¦ 16 10 ¦ 8	26 10	F, 1–3
7	8. bekati, var.	26 5	26 7 ¦ 8	6 11	H
	ibid.	26 5	26 7 ¦ 8	6 11	I
	ibid.	26 5	26 7 ¦ 8	13 ¦ 6 10 ¦ 11	S
8	9. sibu maḥu	6 6	3 ¦ 6 9	16 12 ¦ 11	D, 1–2
9	10. tpā-Xontet	16 6	16 9	16 ¦ 26 12 ¦ 11	D, 3–4
10	11. Xontet-ḥrt	20 ¦ 26 6	26 9	6 12	D, 5–6
11	12. Xontet-Xrt	7 ¦ 6 8 ¦ 7	6 10	26 ¦ 16 12	D, 7–8
12	13. ti-Xontet	16 8 ¦ 7	16 10	26 6 ¦ 12	D, 9–10
13	14. sapti-Xonnu	26 7	20 ¦ 26 10	3 ¦ 6 2 ¦ 1	D, 11–12
14	15. ḥr-àb-ua	¦ 6 . ¦ 8	¦ 6 . ¦ 11	¦ 16 . ¦ 1	Sahel
15	16. sasmu	16 8	16 11	26 2 ¦ 1	D, 13–14
16	17. Kenem..	26 8	16 ¦ 26 11	6 2	D, 15–16
17.	18. tpā-smad	26 ¦ 6 5 ¦ 9	6 11 ¦ 12	6 ¦ 16 2	D, 17–18
18	19. smad	16 9	16 12	26 . ¦ 2	D, 19

	Decan name	Early rise (Inscript. ¦ theoret.)	Middle rise (Inscript. ¦ theoret.)	Late rise (Inscript. ¦ theoret.)	Place on the board
19	20. (sȧt)	10 ¦ 26 / 9	26 / 2 ¦ 12	. ¦ 6/3	D, 20
20	21. ṭȧȷsi-sart	4 ¦ 6 / 10	3+x ¦ 6 / 1	16 / 2 ¦ 3	M, 21-22
21	22. xr-xpd-sart	16 / 10	16 / 4 ¦ 1	8 ¦ 26 / 2 ¦ 3	M, 23-24
22	23. xu 2	26 / 10	26 / 2 ¦ 1	6 / 4	M, 25-26
23	24. xu 2	6 / 11	13 ¦ 6 / 2	16 / 3 ¦ 4	M, 27-28
24	25. bibi	16 / 11	6 ¦ 16 / 2	26 / 3 ¦ 4	M, 29-30
25	26. xont ḥr	26 / 11	26 / 2	6 / 5	M, 31-32
26	27. ḥr-ȧb-xont	6 / 12	. ¦ 6/3	16 / 15	M, 33-34
27	28. xont-xr	10 ¦ 16 / 11 ¦ 12	16 ¦ 16 / . ¦ 3	26 / 5	M, 35-36
28	29. (Kod)	26 / 11 ¦ 12	. ¦ 26 / 3	6 / 6	M, 37-38
29	30. šesbi šesbi	6 / 1	6 / 4	16 / 6	M, 39-40
30	31. xa	16 / 1	16 / 4	16 ¦ 26 / 6	M, 41-42
31	32. ārel	16 ¦ 26 / 1	16 ¦ 26 / 4	6 / 7	M, 43-44
32	33. ȧler saḥ	6 / 2	. ¦ 6/5	16 / 7	M, 45-46
33	34. uārel-xr	16 / 2	16 / 5	26 / 7	M, 47-48
34	35/36 ȧmn xr saḥ	23 ¦ 26 / 2	16 ¦ 26 / 5	16 ¦ 6 / 8	M, 49-50
35	38. sopdet, Sothis	16 ¦ 6 / 2 ¦ 3	16 ¦ 6 / 6	16 / 8	M, 50-51
36	1. ȧpā-kenmut oder šibu genannt	16 / 3	26 13 ¦ 16 / 6	26 / 7 ¦ 8	m, 51-52

The other inscriptions on the table that were not previously displayed are in the following series:

I.

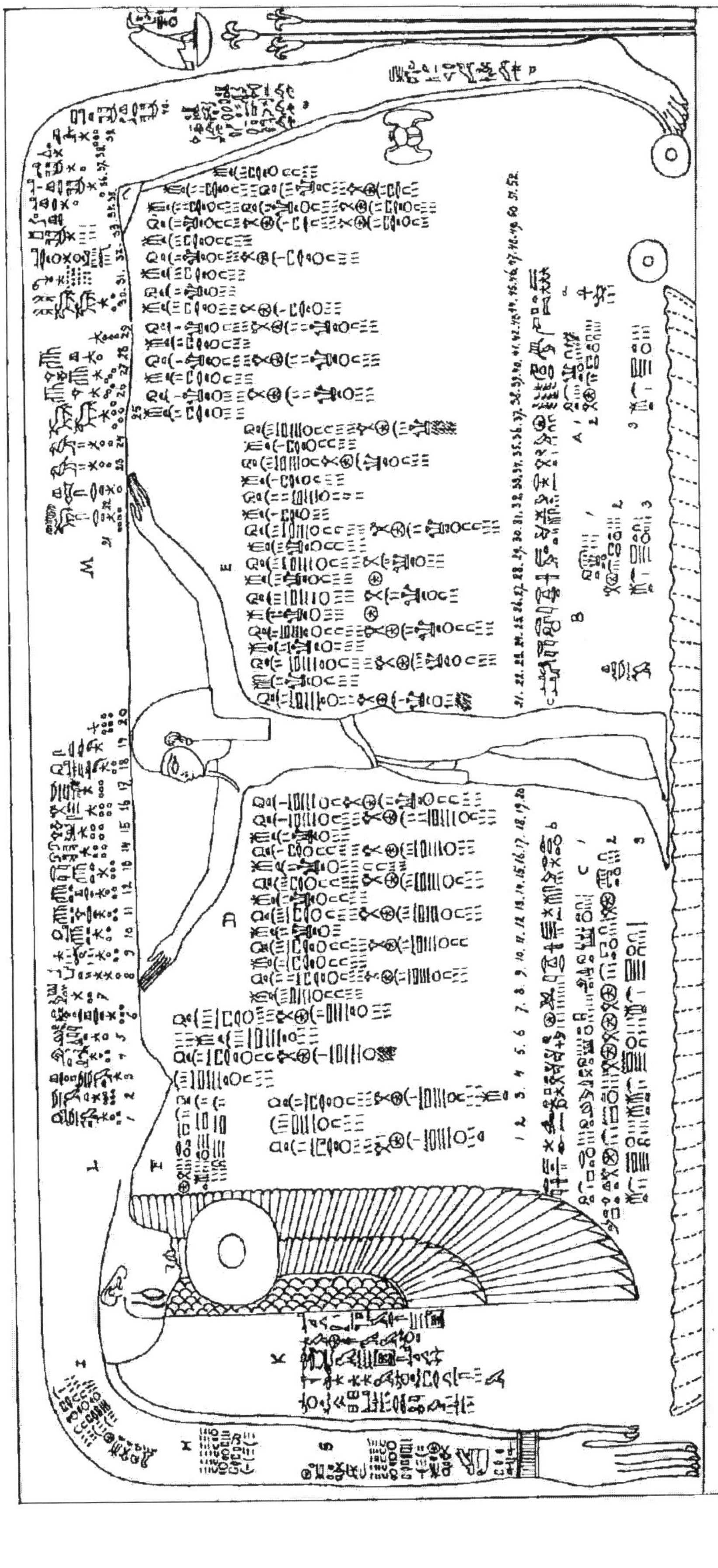

The Decan - Rising according to the representation in the Tomb of King Ramses IV

I. Text on the shin of the female celestial figure (*goddess* *nut*) next to the image of the winged beetle soaring upwards (symbolizing the emerging sun): *su sa er ta Xat-mas* "he (i.e. the sun god) goes to earth, where his birth appears." Above this is a three-line text with the following content: "(1)" He leads the ....?..... to his horn, he is on the back, (2) he breaks through the loins of his mother, the sky goddess *Nut*: (3) he lets himself ascend into heaven.

II. Five-line inscription (K) on the western side of the sky goddess next to the image of the setting sun disk , the content: (1) "There enters the majesty of this god (the sun) to the depth (to be read instead of ?) of the waters of the horizon; (2) the stars of the western lower hemisphere (*duat*) are (3) behind his bark on the waters of the horizon. These (4) wandering stars (*sekod*) behind him are let in and they rise (5) behind him and they turn to their cities (places, stations)."

Remarks. The word hat-θet-taui[61], which I have translated as "*horizon*", literally means "house or space which touches two areas at the same time and separates them from each other." (see my *Geogr.*[62] Lex. p. 983 ff.) In the astronomical sense, hardly any other conception than that of the horizon seems admissible. The term *sekod*, "wandering, circling" for the decan's pictures is to be added to the other names recorded above on page 102. The origin of the name *sekod* is easily recognized from the examples given in *Wort.*[63] IV, see 1781 lin. 9. The *demau*, or *cities* of the decans, evidently refer to the theater in the astronomical sky. It often appears in the inscriptions as a *geographical* area of lands, surrounded by seas and intersected by rivers and canals, covered with cities (,), houses (,), to which special gates ( ) allowed entry, and divided into districts (sapt), which corresponded to the nomes of Egypt, only differing numerically by 36 from them. The latter automatically recalls the number of 36 decans. According to the representations and inscriptions of the king's tombs of Thebes (see Ch. *ND*[64] II, p. 640 ff.) There was a *uat-ura-abti* "eastern sea" (p. 641), a *uat-ura mahtet* "northern sea" (p. 658) and a *uat-ura amenti* "western sea" (p. 682) in the sky. Hence the talk of "the waters and the lands of the northern light side (of the sky) above the constellation of the great bear" (see above p. 94, line 2). The countries of celestial geography include, for example, *Punet* (Ch. *ND* II, 641, 650), and (p. uθenet 641),

---

[61] Translator note: *hat-θet-taui*

[62] Translator note: most likely *Geographische Inschriften altägyptischer Denkmäler*

[63] Translator note: most likely *Worterbuch*; i.e., dictionary.

[64] Translator note: Champollion, *Monuments de l'Égypte et de la Nubie. Notices descriptives.*

kenemti-hr (p. 650; and *ta-nutar-t mahti* "the northern land of God" (p. 658) and other *names*, those of terrestrial geography the images of the gods of the southland and northland (i.e. Upper and Lower Egypt) come to you in the nomes, you go where they are as a perfect soul, you do what you please in heaven, you are among the constellations of the *36 decans*" (*bekau*. See also l.l. pl. 11 Lin. 11).

III. *The long inscription 6-c, cut into 2 parts through the legs of the heavenly bearer, above the texts A.B.C.* In its current version, it suffers from obvious errors on the original, which do not allow a perfectly clear understanding of its content. The attempt of Dr. Gensler (*Ztschft.* 1872, p. 62 ff.) to decipher it, to establish the correct reading, seems to me not to have succeeded in ancient times, especially since the publication of the inscription known to him (in Champollion, Rosellini and Lepsius) suffers from obvious errors and omissions of the modern copists. The end of the inscription is not misunderstood: *Xr au pu pet em sibu* "now that is namely the expansion of the sky towards the [decan] stars." The division into three parallel parts of the preceding text is just as obvious:

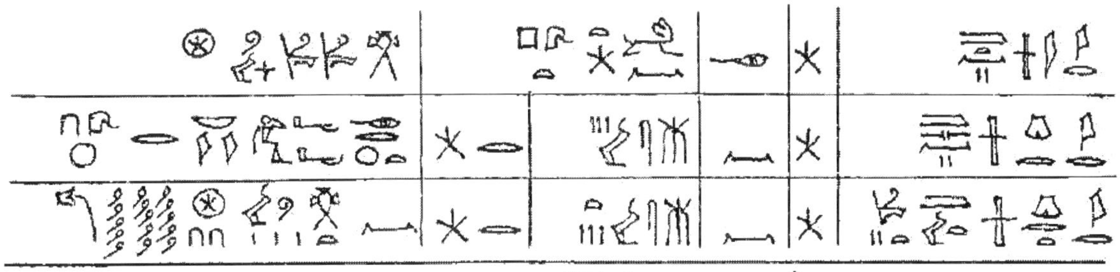

The determination of the interval (, (sic), *amtu*) of (, *en*) two (II, *son*, i.e. stars) after the early rising (*unet tepet*, ie. the first hour of the night), after the midnight rising (*sa-nen amuu duat*, *satuu duat*) and after the late rising (*masu*, *masut*) is clear. Likewise, Dr. Gensler not wrongly suspected that the group refers to the number 20 + 15 = 35 (incorrectly instead of 20 + 16 = 36) of the 36 rising epochs.

The denotation of , var , *duat*, in the sense of "*lower hemisphere*" is indubitable from a long series of instructive examples. I quote the following, which is repeated on monuments of older and more recent dates in the same way and, as it were, forms the introduction to the lists of the daily hours (see below the *calendar in full daily hours*):

"The emergence (i.e. the sun) from the lower hemisphere, the union with the [sunships] *Ma-ad*, the journey on the celestial ocean towards the hour of the day [namely:] *peter-nofiru-nebes*, the emergence in the form of the god *Xoper*, the swinging up to the sphere of light, the entry into the gate [in the east of the sky], the stepping forward with great effort (literally: work), the rising [of the sun] from the opening of the two door panels of the light sphere against the hour of the day [called:] *seXa-nofiru-ra*" [i.e., the first hour of the day].

The List of Decades in the Louvre

The second monument mentioned above, p. 134, line 4, with details which refer only to the 36 ten-day weeks or decades of the Egyptian year, but which are necessarily in close connection with the epochs of the decans, is currently in the collections of the Louvre (D.37). It has been published by Mr. *Pierret* in his *Recueil d'Inscriptions inédites du Musée Égyptien du Louvre,* vol. II, page 73 under the title "*Calendrier*" without any other detailed explanations. Although the monument has only been preserved in a fragmentary state (1/6 of the whole), the layout of the representations and inscriptions can still be seen from the last restoration in great detail. The images and the related texts are repeated in the same way and are arranged in two sections, an upper A and a lower B. Of the general inscription above A, only the following passage has been preserved, the beginning and the end are missing:

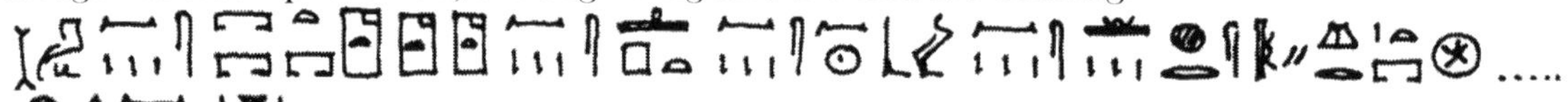

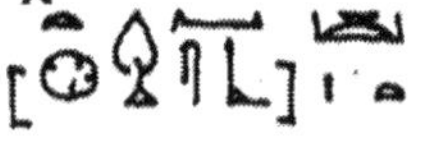 "The lower hemisphere includes its existing rule of its rising and its setting. Their houses are in the the city of *At-[nebeset]*... The recurring pronoun "their," Egyptian , *sen*, can only refer to the 36 decans' stars in context with special consideration of the expressions: *rising* and *setting*, to which 36 decan stars refer. The Egyptian word  *seXer*, rendered by me by "*existing rule*," recurs in the same astronomical sense in the Decree of Canopus. In the hieroglyphic part of the same (line 20) it says once:  *ma seXer uon petel* (sic) *smen hirs* i.e. according to the existing rules (or laws) on which the sky (i.e. the astronomical phenomena in the sky) is based. For its part, the Greek text has for it (line 41): κατὰ τὴν νῦν οὖσαν κα[τασ]τάσιν τοῦ κόσμου "According to the order of the world that now exists," while the demotic translation of the same passage (line 40) reads: 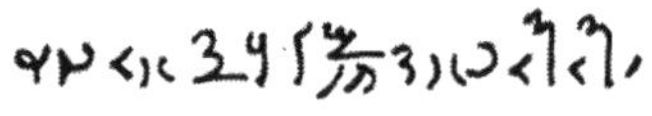 "According to the scheme on which heaven is founded."

Of the general heading to be referred to in Section B, only the following groups have survived   .... "…. his hidden healing (?). It is they who cause the hurricanes, it is they who bring about the showers; the preventers of the sun's rays. They dwell for a while, concealing…. "

The place  *At-nebeset* (i.e. place of the *nebes*-tree)," sometimes also 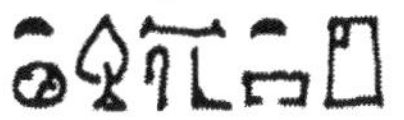 *hat-nebeset,* and also written only  *hat-nebes,* which is frequently mentioned in the inscriptions of the monument, located in the lower Egyptian region called nomus Arabia by the ancients , on the easternmost side of the delta region (compare Br. GW,[65] 333 ff.) A special local form of the god  *su* was worshiped there under the name  *soped,* after which the same city, the metropolis of the Arabic nomus, had the holy name  *Pi-soped* "city of the god Soped." The same place is to be understood in the cuneiform inscriptions under the name *Pesept, Pi-sap-tu,* and without *Pi: Saptu,* while today the old designation of the same has survived in the modern *Saft.* This is still the name of a village located in the southeast direction of Zagazig (i.e. Bubastus). See my remarks in the *Ztschft* ,1882, p. 15 ff. It is evident from this that our monument comes with unquestionable certainty from ancient *Pi-soped.*

The monument, preserved in its entirety, consisted of 2x18 or 36 fields, each of which is arranged according to the following scheme with three different pictorial representations of the (warlike) god Soped:

[65] Translator note: Brugsch. *Die ägyptische Gräberwelt*

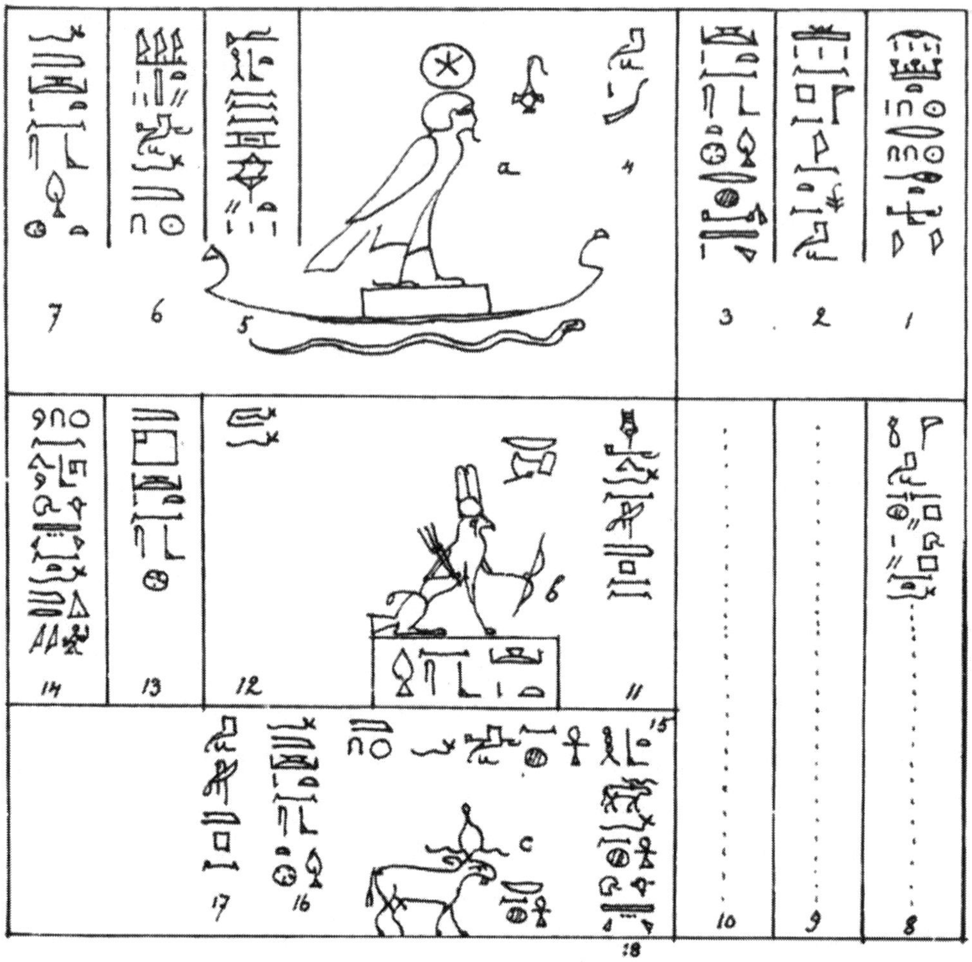

Translation and description.

"On x/y up to day x + 9 [in the above example: 11 *Choiak* up to 20] sacrificial offerings are made (1) to this god by the king in (2) the city of *At-nebeset* in order to protect the country (3) from damage (4)"

a. Image of the god next to it  *neser* "fire flame."
"What is needed in water, winds (5) and fields lies in his hand in his decade (6) in the city of *Al-nebeset*" (7). He is the great god from the beginning [the following varying formulas describe the anger and act of revenge of god against everything hostile 8-10]. Its position is according to this representation (11)

b. Image of the god called *neb Xer* "Lord of the fight," tuner of the pedestal: the city of *At-nebes*.
"He emerges (12) from the house [in the astronomical sense!] of the city of *At-nebes* (13). Throughout his decade, after his mission on earth, it is he who causes death (14). Whatever life hath lies in his hands throughout his decade (15) in the city of *At-nebeset* (16) according to this illustration (17).

c. Image of god called *neb-anX* "Lord of Life." Before: "his living ram on earth" (18).

It is evident from these inscriptions, it seems to me, that the deity of *At-nebeset* (in his threefold conception of flame, lord of battle and lord of life) simultaneously represented the decan of the respective decade after those observed in the nomus Arabia noted epochs of the rise. Apparently that represents that line 13 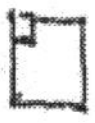 *hat*, "*house*" has the scope of observation. The decades preserved on the monument and their relationship to one another emerges most clearly in the following supplementary overview.

The decade series on the memorial in the Louvre

Section A. series	Monument	Decade: Starts	Decade: Ends
1	·	1/1	10/1
2	·	11/1	20/1
3	·	21/1	30/1
4	·	1/2	10/2
5	·	11/2	20/2
6	·	21/2	30/2
7	·	1/3	10/3
8	·	11/3	20/3
9	·	21/3	30/3
10	·	1/4	10/4
11		11/4	20/4
12		21/4	30/4
13		1/5	10/5
14	·	11/5	20/5
15	·	21/5	30/5
16	·	1/6	10/6
17	·	11/6	20/6
18	·	21/6	30/6

Section B. series	Monument	Decade: Starts	Decade: Ends
19	·	1/7	10/7
20	·	11/7	20/7
21	·	21/7	30/7
22	·	1/8	10/8
23		11/8	20/8
24	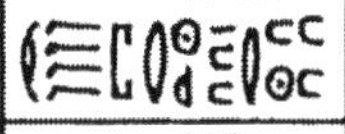	21/8	30/8
25	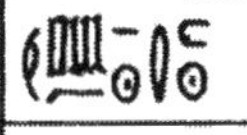	1/9	10/9
26		11/9	20/9
27	·	21/9	30/9
28	·	1/10	10/10
29	·	11/10	20/10
30	·	21/10	30/10
31	·	1/11	10/11
32	·	11/11	20/11
33	·	21/11	30/11
34	·	1/12	10/12
35	·	11/12	20/12
36	·	21/12	30/12

The Timetables in the Tombs of Kings Ramses XI and Ramses IX at Thebes

In the tombs of kings *Ramses VI* and *Ramses IX*, identical, albeit deviating and incorrect copies of a correlated, hieratically composed original astronomical document were first discussed by *Champollion* (*Lettres ecr. d'Egypte,*[66] p. 239), albeit misinterpreted, later recognized by *Lepsius* (introduction[67] p. 110 ff.) and used scientifically and finally by *Dr. Gensler* (*Die Thebanischen Tafeln stündlicher Sternaufgänge,* Leipzig, 1872) checked and explained from an astronomical point of view. Despite this meritorious preliminary work, they still await their complete resolution by computational astronomy. Having endeavored to correct the errors and unforgivable mistakes in the work of my predecessors from the philological point of view, I present the result of my investigations in the following overview.

Like Lepsius first lines have shown, the indexes, arranged evenly in 24, refer to the rising of a series of stars at the onset of night and in the 12 hours of the night that follow thereafter, in the 12 monthly and 12 semi-monthly (the latter indicated by  "day 16 . . . 15") epochs of the Egyptian year. The individual views are based on their local position in relation to eight (not seven, as Lepsius assumes) body parts of a crouching male figure, with the face toward the viewer, the drawing of which is intended to be distributed over 8x13 or 104 square fields. The figure and grid of the sky man are separated from each other in the drawing for each monthly and semi-monthly timetable, but they necessarily belong together. The positions of the stars are entered in the squares at the appropriate places and indicated in the accompanying text with the constant formula: "*the star x on* ( *hr*) *this or that entity*." The individual limbs of the man,

[66] Translator note: *Lettres écrites d'Égypte et de Nubie en 1828 et 1829.*
[67] Translator note: This probably refers to *Die Chronologie der Aegypter.*

together with the designation I have chosen for the sake of brevity, are as follows:

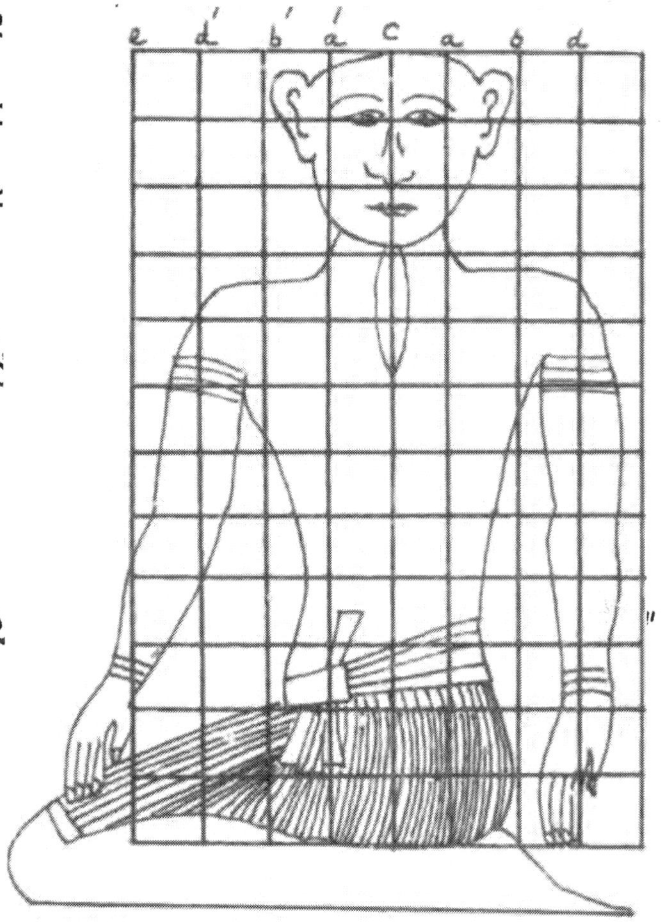

[For the meaning of <u>*masXen*</u>, see above, p. 94].
The drawing below shows the image of the crouching man inside the square grid, based on Dr. Gensler's design below:

The stars, recorded according to their hourly rise, belong to smaller and larger constellations. Depending on their size and extent in the sky, a more or less large number of the special parts of the same are listed, which appear as limbs in the pictures of living beings (humans or animals).

By taking the constellation of Sirius (Sothis) as the starting point, the realms of the constellations are clearly arranged in the following way:

I. 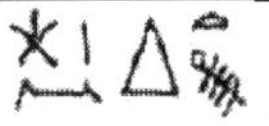 <u>*sib en sopet*</u> "Star of the Triangle," Sothis-Sirius

1. 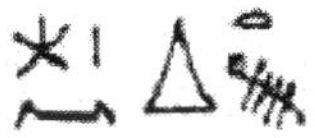 "Star of the triangle."

2. 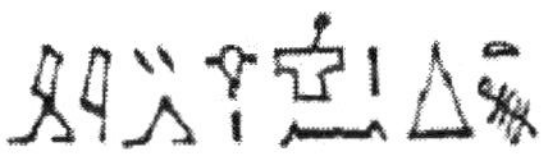 <u>*i hr sa en sopdet*</u> "what follows the triangle."

II.  *sibui* "the double star."

3. 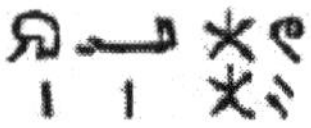 *spa-sibui*,  *tp-sibui* "tip (head) of the double star"

4.  *sibui*, "the double star"

III. 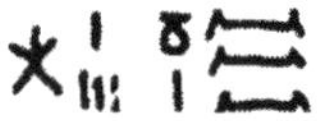 *sibu bu mu* "the stars of the water."

5. 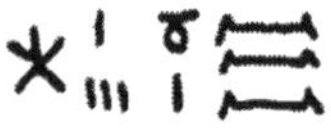 *sibu bu mu* "the stars of the water."

IV. 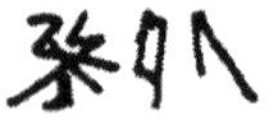 *maa* (var. . *maa*,  *mai* or  ?) "the Louvre."

6.  *tp en maa* "the lion's head."

7. 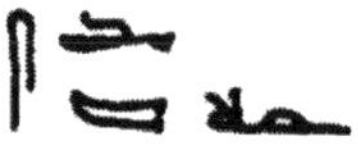 *sedef* "be pregnant"

V.  *sibu asu* "the many stars"

8. 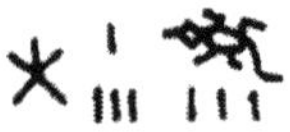 "the many stars, the many heavenly bodies."

VI.  *ta-nofir* "the beautiful boy."

9.  "the beautiful boy."

VII. 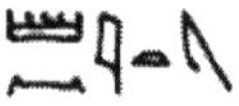 *menat* "the (upright) knife." (see above page 99). Compare 15.

10.  *sesu en hatu menat* "the puller at the beginnings of the knife"

11. 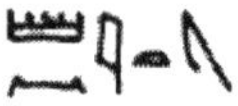 *menat* "the knife."

12. 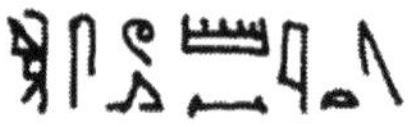 *sesu menat* "the knife puller."

VIII.  *reret* "the female hippopotamus" (see above p. 93 and p. 98, 2).

13.  *todui en reret* "the 2 feet of the Niepferd figure"

14. 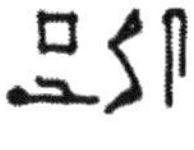 *pades* "her leg. "her leg"

15.  *hri-ab menat* "the mill of the knife." see VII.

16. 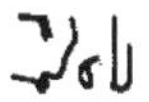 (var 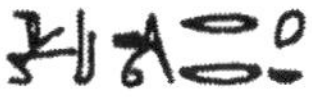 *as en reret*) "your kidney"

17.  *bahes* "her cut (?)"

18.  *Xepdes* "their shame"

19. 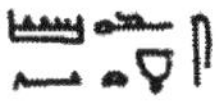 *mendes* "her udder"

[VIII.  "The female hippopotamus"]

20. 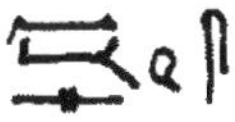 *neses* "their tongue".

21.  *sutis* "your double feather"

IX.  *naXet* "the giant"

22. 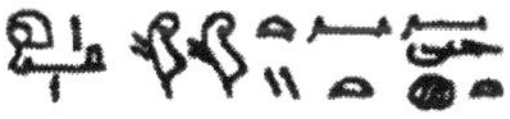 *tpa suti ent naXet* "the tip of the giant's double feather".

23.  *suti ent naXet* "the double feather of the giant".

24. 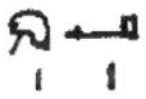 (var. ) *Tpa (tp) ent naXet* "the head of travel."

25.  *nehebef* "his halo."

26.  *Xabulef* "his neck."

27.  *begasetf* "a collar"

28. 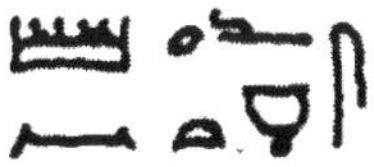 *mendeles* "his chest."

29. 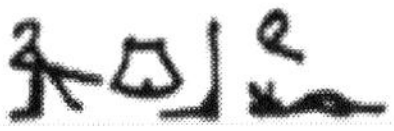 *agebef* "his knee"

30.  *setehetef* "his shin."

31.  *patef* "his foot."

32.  *sebeqef* "the sole of his foot."

33.  *peXes* "his curling board

34. 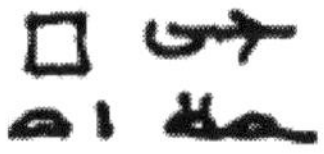 *patef* "his base (?)."

X. 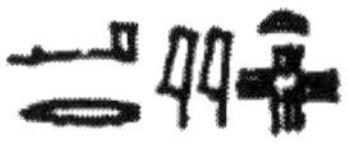 *arit* "the star Arit." see p. 40-41

35. *arit* "Arit,"

XI.  *aped* "the bird" (goose?)

36. 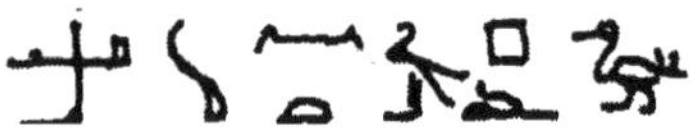 *ab ent aped* "the hood of the bird."

37.  *tp en aped* "the head of the bird."

38.  *keftutef* "his hindquarters."

XII.  *sib en Xau* "the thousand stars"

39.  *sib en Xau* "the thousand stars"

40. 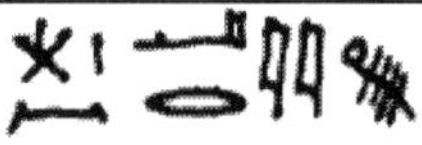 *sib en ari* (var. 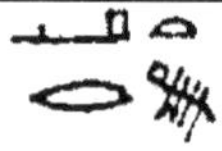 *Aret*) "the star *Arit*." See No. X.

41.  *sib en se-arit* "the star *Se-arit*."

XII. 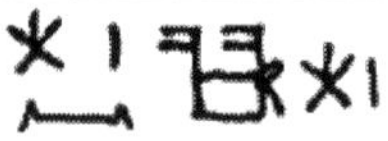 *sib en sah* "the star of Orion."

42. 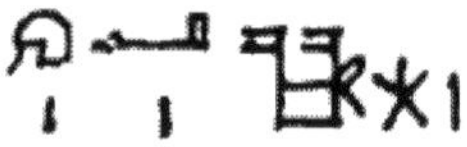 *tpa sah* "the tip of Orion."

43. 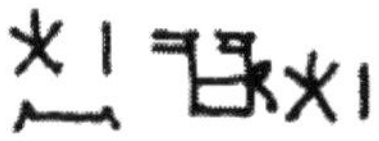 *sib en sah* "the star of Orion."

The tables that now follow contain an exact description of the Theban hourly tables in a tabular overview, whereby I have chosen the Sothis star as the starting point, the rising of which first appears on 16-15 Thoth. *O* denotes the *entry of the night*. The numbers from *1* to *12* the individual *hours of the night*. Because the limbs of the Heavenly Man are designated by c, a, b, d, e, a', b', d', I refer to my remark above. The letters (a), (a'), (b), (b') etc. enclosed in brackets signify *variants*. Understanding the table shown above does not present the slightest difficulty. Above all, one will gain the conviction that in the horizontal rows the numbers of the night hours are listed in regular order. The individual constellations are differentiated according to their sequence by the numbers added to the German names, whereby the constellation of Sothis or Sirius forms the starting point for the count.

Month	Day	1. Star of Sothis	2. what follows Sothis	3. Head of the double star	- Point of the double star	4. the double star	5. the star of the water	6. the head of the lion	7. his tail	8. lots of stars	9. the beautiful boy
Thoth	16–15	12, d (d')									
Phaophi	1	11, d'			12, c						
ibid.	16–15		10, d (d')		11, d'		12, c				
Athyr	1		9, a		10, c		11, c	12, c			
ibid.	16–15		8, a		9, c		10, c	11, c	12, c		
Choiak	1		6, c'		7, d'	8, a	9, e	10, d	11, d	12, d'	
ibid.	16–15	(5, d)	5, d			6, d	7, d'	8, d'	9, d'	10, c	11, c
Tybi	1		4, d'			5, b'	6, b'	7, d'	8, c	9, c	
ibid.	16–15			3, c		4, b'	5, c	6, c	7, a	8, c	
Mechir	1	1, b				2, b'	3, d'	4, b'		5, c	6, c
ibid.	16–15		1, d'			2, a		3, b'	4, d'	5, c	6, c
Phamenoth	1	0, b			1			2, b'	3, c	4, c	5, c
ibid.	16–15				1				2, c	3, c	4, c
Pharmuthi	1							0, c	1, d'	2, a	3, a
ibid.	16–15									1, d'	2, a

Month	Day	10. the puller at the beginning of the knife	11. the knife	12. the puller of the knife	13. the 2 feet of the hippopotamus figure	14. her leg	15. the middle of the knife	16. her kidney	17. her cut (?)	18. her pubic	19. her udder
Choiak	16-15	12, c									
Tybi	1	10..	11, á	12, á							
ibid.	16-15	9, d́	10, á	11, d́	12, d́						
Mechir	1	7, c	8, b́	9, c	10, d́	(10)	11, á			12, c	
ibid.	16-15	7, c	8, b′	9, c	10, d́		11, á			12, c	
Phamenoth	1	6, a		7, a	8, á	9, c	10, c			11, c	12, b
ibid.	16-15	5, a	(c) 6, a	7, a	8, á	9, c				10, c	11, b
Pharmuthi	1	4, á		5, a	6, á	7, c	8, c			9, c	10, b
ibid.	16-15	(á) 3, a	0, á	(a) 4, á	5, á	6. =			7, c	8, d	9, d
Pachon	1	1, á	0, b 2, á	3, á	4, c	5, c			6, b	7, d	8, b
ibid.	16-15	0, á	1, a	2, á	3, c	4, c			5, d	6 d́	(b) 7, b′
Payni	1		0, c		1, c	2, á		3, c		4, c	5, á
ibid.	16-15				0, c	1, c		2, c		3, c	4, a
Epiphi	1							0, á		1, c	2, a
ibid.	16-15									0, c	1, c
Mesori	1										0, c

Month	Day	20. her tongue	21. her double feather	22. the tip of the giant's double feather	23. the double feather of the giant	24. the head of the giant	25. his throat	26. his neck	27. his collar	28. his chest	29. his knee
Thamenoth	16–15	12, d									
Pharmuthi	1	11 d (b)		12, b							
ibid.	16–15	10, d		11 …	12, b						
Pachon	1		9, b	10 …	11, b	12 b′					
ibid.	16–15		8, b	9, b′	10, b′	11, c		12, d			
Payni	1	6, á	7, b	8, b	9, c			10, c		11, c	12, c
ibid.	16–15	5, a	6, b	7, b	(7, b′)	8, c		9, c		10, d	11, b
Epiphi	1	3, c	4, c	5, c	6, á		7, c			8, c	9, c
ibid.	16–15	2, c	3, c	4, c	5 b (b′)			6, c		7, c	8, c
Mesori	1	1, c	2, c	3, c	4, a			5, c		6, c	7, c
ibid.	16–15	0, c	1, c	2, c	3, a			4, c		5, c	6, c
Thot	1				0 …	1 (d)	2 a (c)		3, c		
ibid.	16–15					0 (a)	1, a		2, c		
Phaophi	1						0, c		1, a (c)		
ibid.	16–15								0, c		

Month	Day	30. his shin	31. his foot	32. his sole	33. his base board	34. his pedestal	35 the star ārit	36. the hood of the bird	37. the head of the bird	38. his posterior	39. the thousand star.
Payni	16-15		12,c								
Epiphi	1		10,d́	11,c	12,c						
ibid.	16-15		9,c	10,c	11,c		[....]				
Mesori	1		8,c	9,c	10,c		[....]	[....]			
ibid.	16-15		7,c	8,c	9,c.		[....]	[....]	[....]		
Thoth	1	{5,d {4				(a) 5,c	(c) 6 a		7,c	(á) 8,c	9,c
ibid.	16-15	3,d				4,c	5,a		6,c	7,c	8,c
Thaophi	1	2,c				3,c	4,c		5,a	6,c	7 c
ibid.	16-15	(1,c)	1 c			2,c	3,a	(á) 4,a		5,c	6,a
Athyr	1		0,c			1,c	2,a		3,a	4,c	5,c
ibid.	16-15					0,c	1,a		2,c	3,c	4,c
Choiak	1						0,d́		1,c	2,a	3,d
ibid.	16-15								0,c	1,c	2,b
Tybi	1									0,c	

Month - Day	1/1	16-15/1	1/2	16-15/2	1/3	16-15/3	1/4	16-15/4	1/5	16-15/5	1/6	16-15/6
40. Star of Ārit	10,a	9,a			6,c	5,d́						
41. Star of Se-ārit	(10,c)		8,d́	7,a		5,a	4,d	3,b	1,d́			
42. Top of Orion	11,c	10,c	9,á	8,a	7,c	6,c			2,c	0,ĉ		
43. Star of Orion	12,c (12,d)	11,c	10,d́	9,d	8,d́	7,a	5,d́	4,c	3,a	1,d́	0,d́	0,d́

Translator's Appendices

(All page references in the appendix refer to this book.)

Appendix A. The Baboon

From page 68

In the following exerpt Brugsch discusses this line of text from the courtyard of the Temple of Edfu:

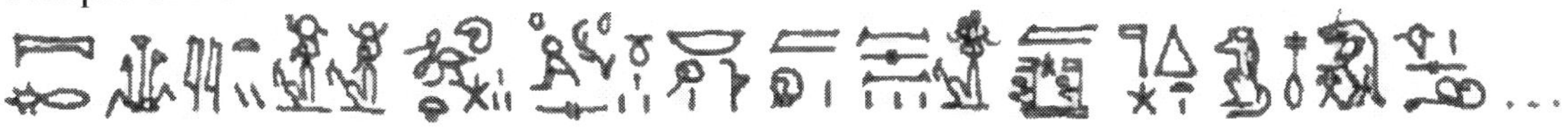

> "the sky (or "the roof" of the temple, the inner ceiling of the same) carries the two light-bringers sun and moon, the decans are behind them, the lord of the beginning of the year is first for them. Osiris as Orion, the divine Sothis, the good (or beautiful) monkey, the hippopotamus ----."

Characters appearing in the text are astronomical:

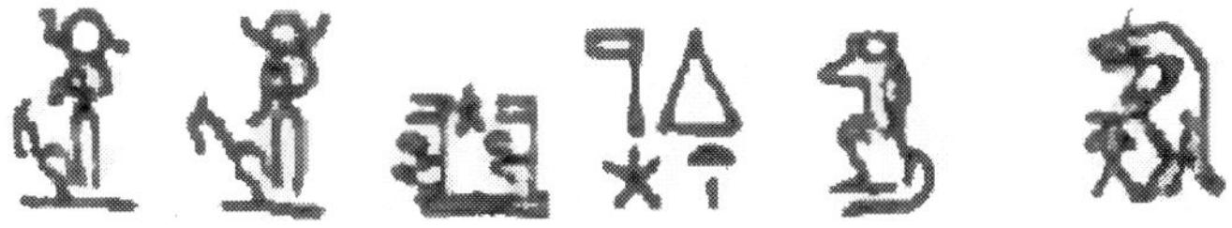

Sun Moon Osiris Sothis Baboon Hippopotamus

There was no "monkey" constellation, so the ape figure should be "baboon," as illustrated in the following 2 images.

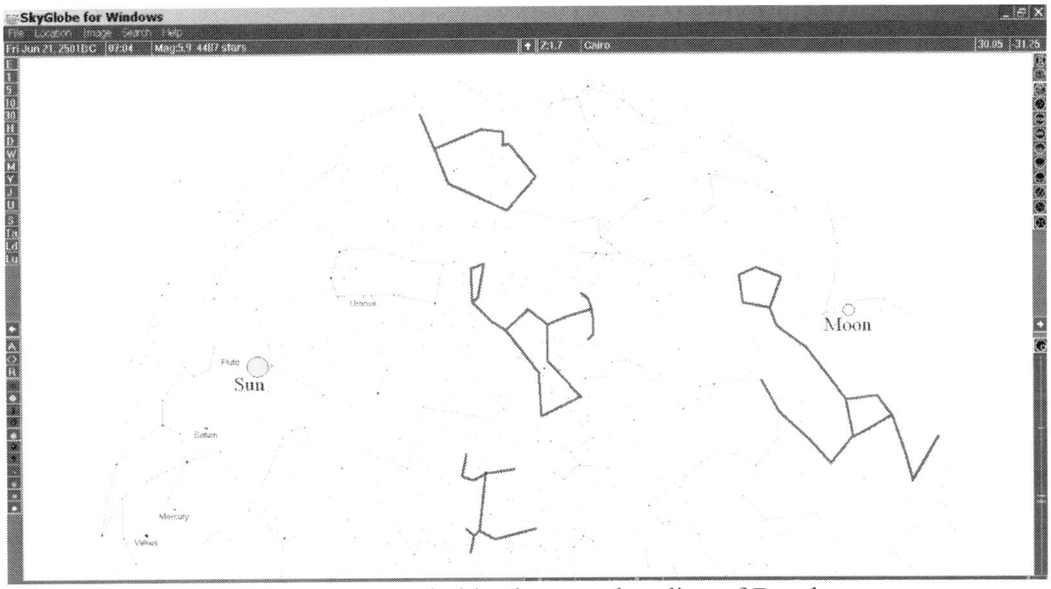

Star images recorded in the round zodiac of Dendera

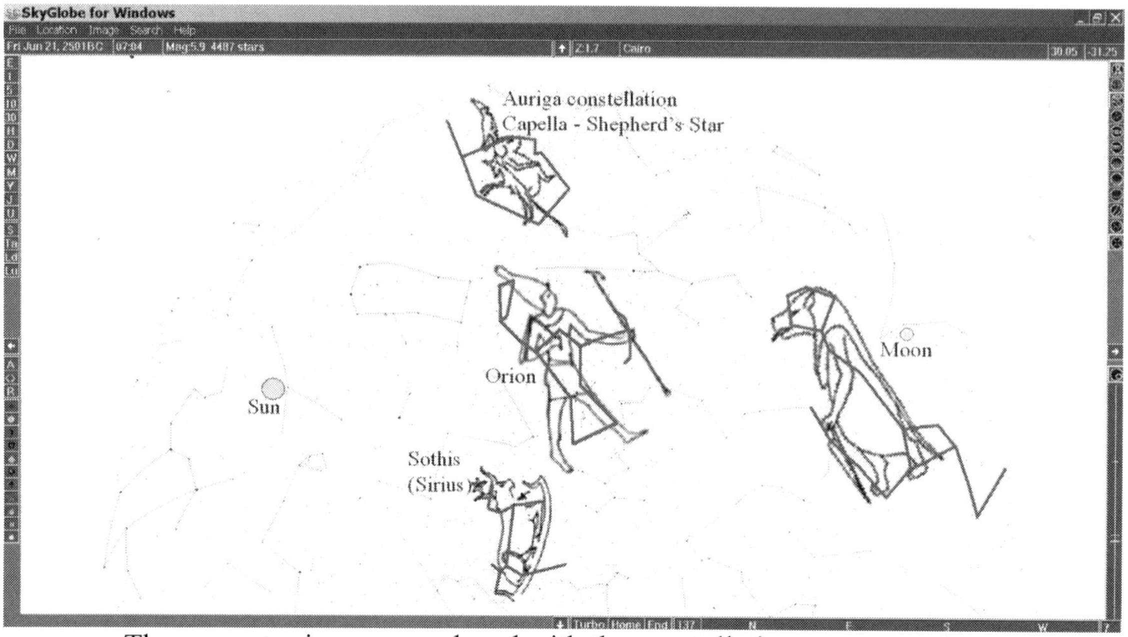

The same star images overlayed with the constellation representations.

Appendix B. The Constellation of the 2 Turtles

(see page 87)

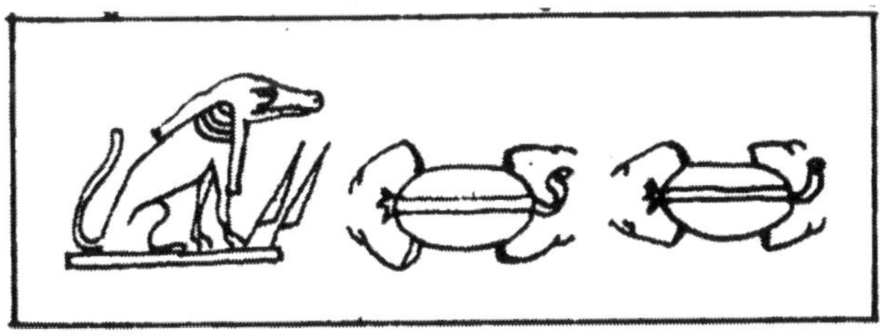

Below is an excerpt from Brugsch's comments on the constellation of the 2 turtles:

> "The above panel [p. 85] under groups A.B.C.1 name a constellation Sethu (A.B.) or Setu (C.), which is to be translated with consideration of the following signs: "the 2 turtles" or "the turtles." The same is not foreign to the Greco-Roman era, as illustrations and star lists prove. On the sarchophagus of the Theban Priest Hetar, the two turtles appear in front of a seated lion with the head of a hippopotamus or crocodile, 2 swords at his forefeet:"

When the positions of the constellations (*below*) are compared with Brugsch's drawing (*above*), it can be realized the star of the turtle meant one of the stars of Cancer, which in those days was known as the constellation of the tortoise or turtle.

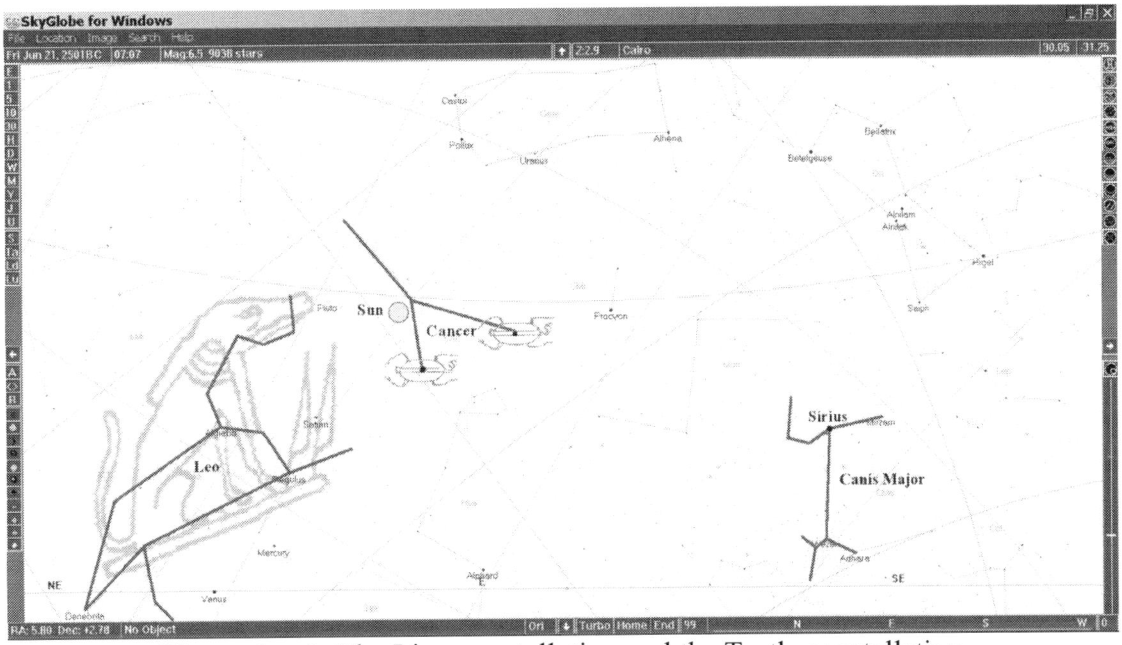

Illustration 1: The Lion constellation and the Turtle constellation.
"The star of the *turtle*  stands between the Sun and Sirius."

Further on page 86 Brugsch says:

> "While in the illustration given above on page [52] (see E) the star of the turtle  stands between the Sun and Sirius. It is consistent with this position that in the more recent decan lists (see below) a constellation  is introduced into the realm of the decans…"

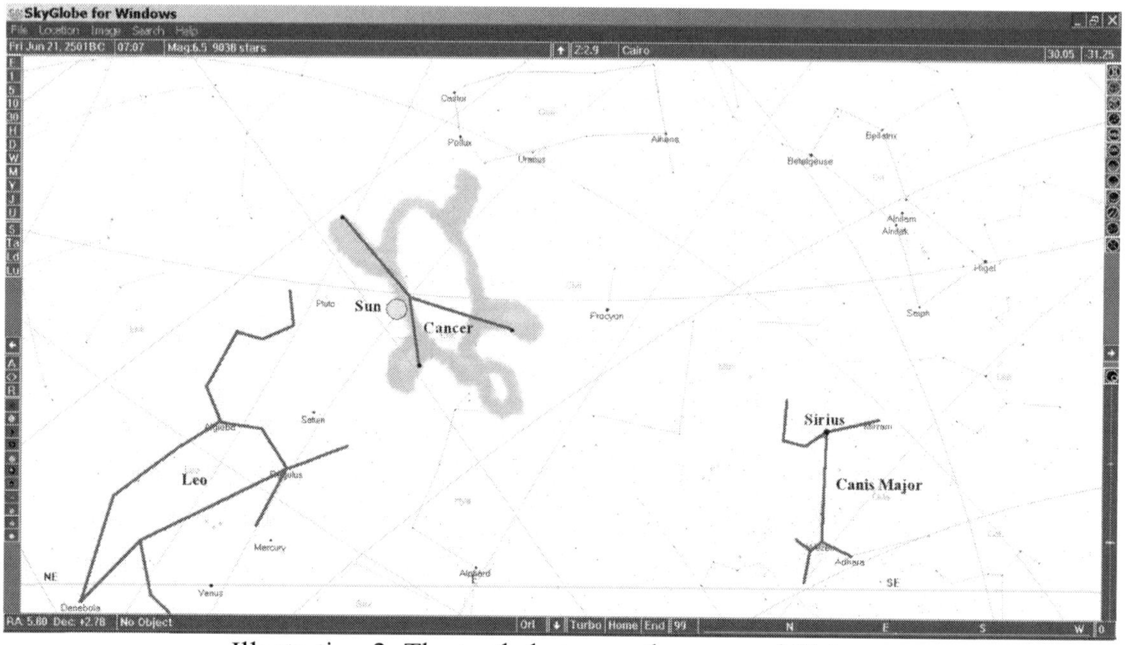

Illustration 2. The turtle between the sun and Sirius

The Greeks thought of the constellation as a tortoise or turtle, but to the ancient Egyptians it was a scarabus beetle.

Ptolemaic Tomb at Atfieh[68]
150 B.C.

To the east of the second pyramid of Giza, a burial discovered by Georges Daressy in February 1902 and dating from the 30th dynasty depicted constellations from 2501 B.C., the time of the positions of the stars, planets, and constellations as seen on the ceiling of the temple at Dendera.

The northern constellations are found towards the north side of the vault, hence under the feet of the great goddess; they occupy a rectangle in width surrounded by protective spirits. The relative disposition of the characters can be indicated as follows:

---

[68] *Annales du Service des antiquités de l'Égypte*, volume III, 1902, page 160. <http://gallica.bnf.fr/ark:/12148/bpt6k5724881p>.

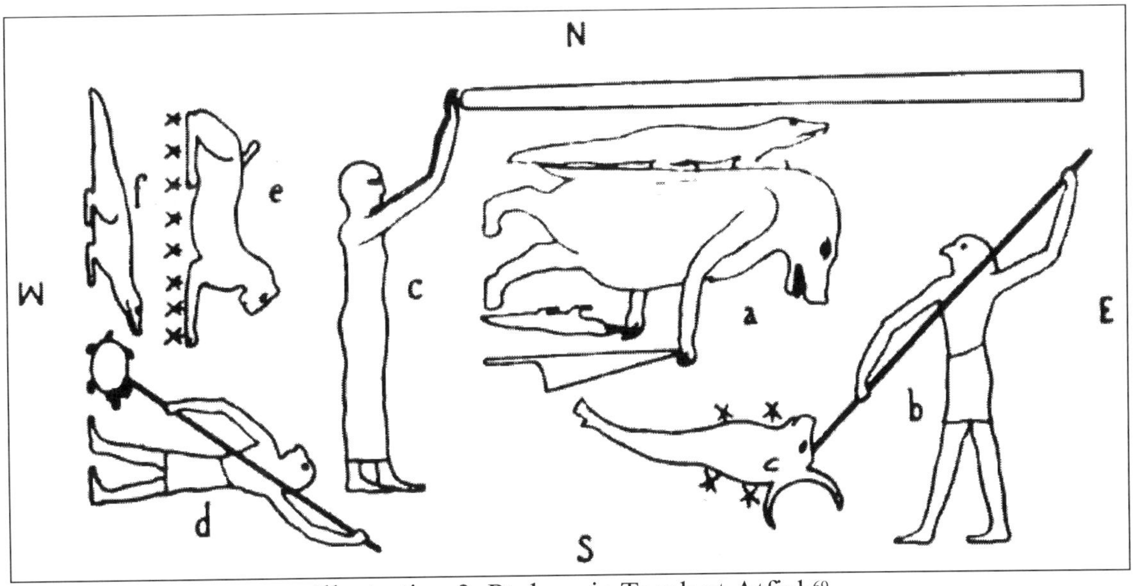

Illustration 3. Ptolemaic Tomb at Atfieh[69]

a. A female hippopotamus, standing on her feet, one of her hands resting — on a large knife, the other on a crocodile standing on her tail. A large crocodile is placed on the back of the hippopotamus. Above the head is the name .

b. A Horus with the head of a hawk placed at the right angle of a hippopotamus pierces with his spear  a thigh ending in the head of an ox 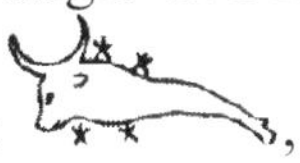, *(Meskh-ti* ) [70] Stars surrounded this figure: only four remain and the name has disappeared.

c. A goddess standing, under the feet of Hippopotamus, facing east; the mutilated name is  (probably for ) Her two arms raised in front of her lean against a long red band, slightly triangular area that passes over the hippopotamus and goes to the eastern end of the painting.

d. Towards the west, a hieracocephalic Horus turned towards the north, pierces a tortoise with his spear. 

e. Facing this Horus, under the goddess Selk, is a lion lying on eight stars, which is called, , Star ('Αστήρ) of Ra.

---

[69] *Annales du Service des antiquités de l'Égypte*, volume III, 1902, page 178. <http://gallica.bnf.fr/ark:/12148/bpt6k5724881p>.
[70] Budge, *Egyptian Hieroglyphic Dictionary*, 326A, *Meskh-ti.*

f. In the northwest corner, below the lion, a large crocodile whose name is mutilated:
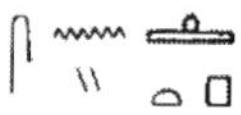

This tomb proves that during the Ptolomaic era, the constellation Cancer was illustrated as a turtle or tortoise. The figure "d" shows the sun, i.e. Horus holding a balancing pole touching the constellation Cancer. This indicates that the sun was in the constellation cancer[71] (see star constellations below). The other hawk headed man with a pole is *An* from the tomb of Seti I. The dotted line indicates the position of the pole held by *An* in the tomb drawing at Afieh. (See *Secrets in the Stars of Ancient Egypt* for other examples from various tombs.)

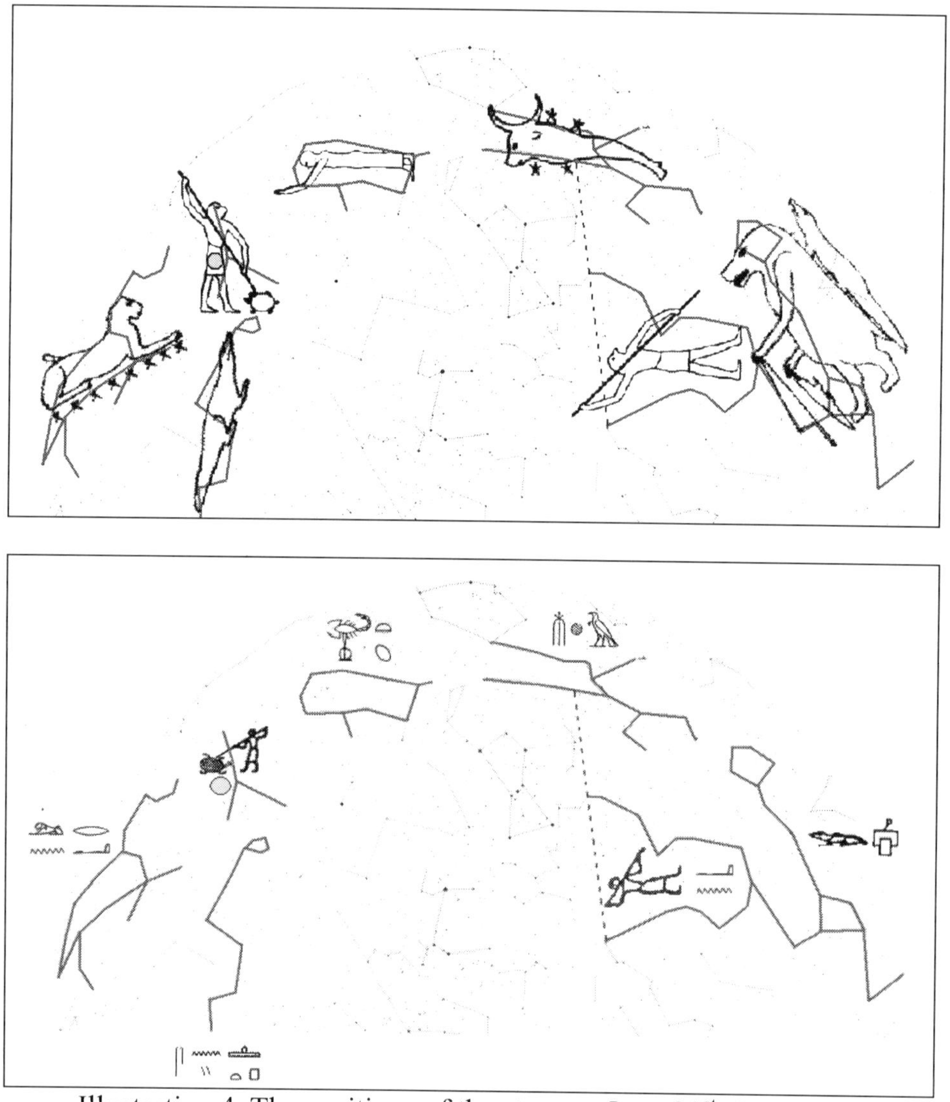

Illustration 4. The positions of the stars on June 21st, 2501 B.C.

[71] June 21st, 2501 B.C.

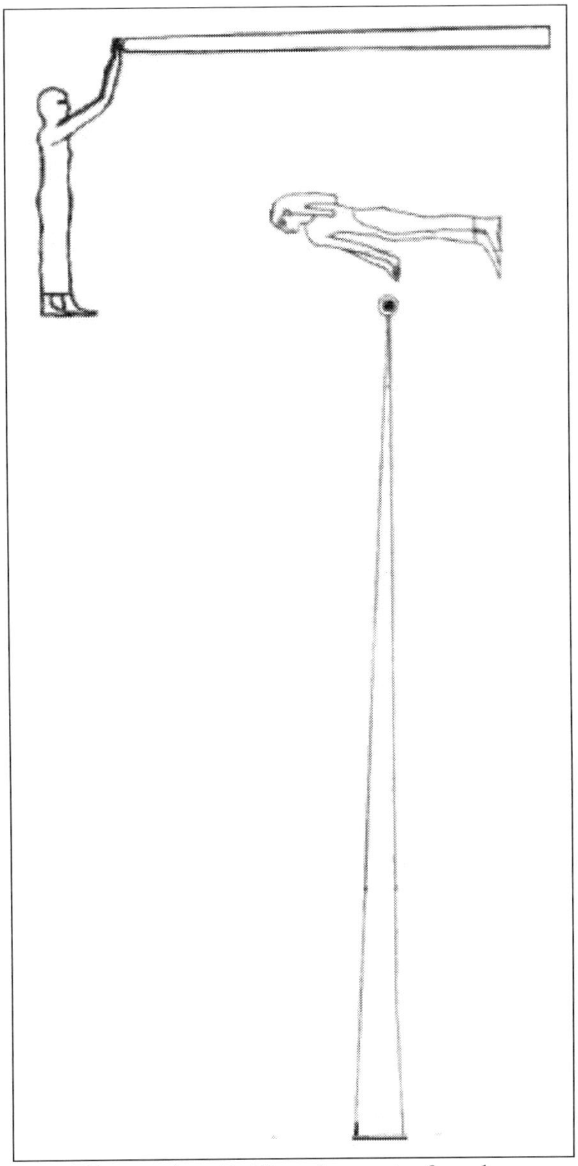

Illustration 5. Hanging set of scales

Each of these two tomb illustrations show a hanging set of scales, representing the pivotal points of the solstices and equinoxes during the apparent orbit of the Sun along the ecliptic path. The top one is from the tomb in Atfieh (see ill. 3, C). The lower example is from the tomb of Senenmut (see ill. 6).

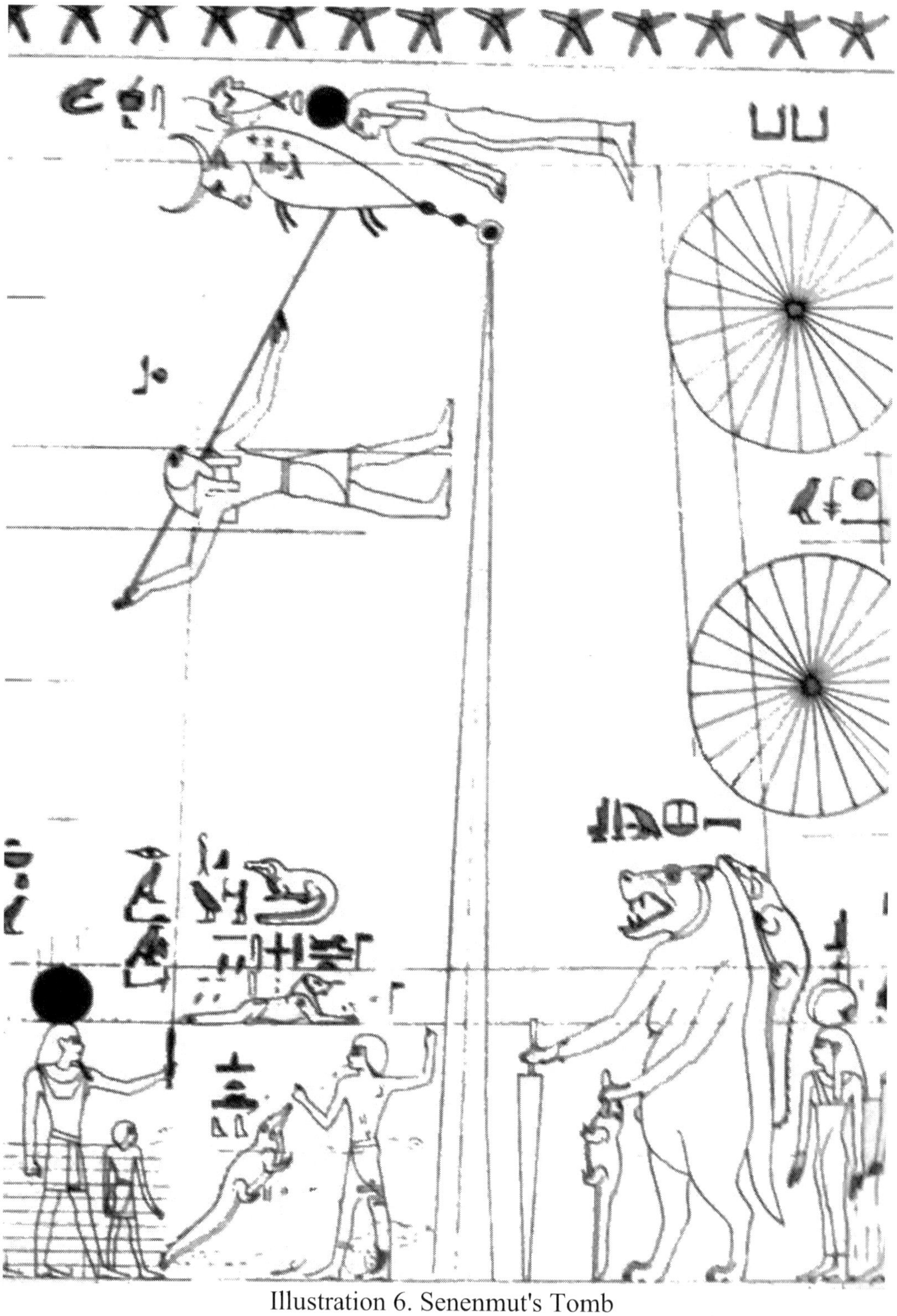

Illustration 6. Senenmut's Tomb

Made in United States
Orlando, FL
04 November 2026

90201198R00098